SpringerBriefs in Finance

SpringerBriefs present concise summaries of cutting-edge research and practical applications across a wide spectrum of fields. Featuring compact volumes of 50 to 125 pages, the series covers a range of content from professional to academic. Typical topics might include: a timely report of state-of-the art analytical techniques, a bridge between new research results, as published in journal articles, and a contextual literature review, a snapshot of a hot or emerging topic, an in-depth case study or clinical example, and a presentation of core concepts that students must understand in order to make independent contributions. SpringerBriefs in Finance showcase emerging theory, empirical research, and practical application in corporate finance, banking, financial management, behavioral finance, financial markets, social and entrepreneurial finance, microfinance, and related fields, from a global author community. Briefs are characterized by fast, global electronic dissemination, standard publishing contracts, standardized manuscript preparation and formatting guidelines, and expedited production schedules.

More information about this series at http://www.springer.com/series/10282

Eline Van der Auwera • Wim Schoutens •
Marco Petracco Giudici • Lucia Alessi

Financial Risk Management for Cryptocurrencies

Springer

Eline Van der Auwera
Department of Finance, Accounting and Tax
KU Leuven
Brussel, Belgium

Wim Schoutens
Department of Mathematics
KU Leuven
Heverlee, Belgium

Marco Petracco Giudici
Joint Research Centre - IPSC - G09
European Commission
Ispra, Italy

Lucia Alessi
Joint Research Center
European Commission
Ispra, Italy

ISSN 2193-1720   ISSN 2193-1739   (electronic)
SpringerBriefs in Finance
ISBN 978-3-030-51092-3   ISBN 978-3-030-51093-0   (eBook)
https://doi.org/10.1007/978-3-030-51093-0

Mathematics Subject Classification: 91G

This Springer imprint is published by the registered company Springer Nature Switzerland AG.
The registered company address is: Gewerbestrasse 11, 6330 Cham, Switzerland

Preface

Cryptocurrencies are digital currencies often embedded on a blockchain technology. More precisely, blockchain technology is the underlying mechanism which ensures that cryptocurrencies can become totally decentralized. In fact, cryptocurrencies are the first successful realization of a decentralized peer-to-peer digital payment system.

Blockchain stores all the transactions from the initiation of the cryptocurrency and ensures its security. The technical details concerning blockchain are described in Chap. 1. It is key to comprehend the underlying technical specifics to understand how cryptocurrencies work and where some of the risks lie. In particular, an important ingredient is the consensus algorithm that dictates which transactions are labelled as valid. In the same chapter, also some of the consensus algorithms in cryptocurrencies are briefly discussed.

Bitcoin is currently the most renowned cryptocurrency. It was "invented" by Satoshi Nakamoto[1] in 2008. After Bitcoin, many other cryptocurrencies have seen daylight, some of them are valuable variations on Bitcoin, implementing different consensus algorithms, allowing additional flexibility, using different encryptions; others are scams. Chapter 2 is dedicated to explain some of the most important cryptocurrencies together with their specific features and characteristics. The different cryptocurrencies offer a broad spectrum of applications and many of them tackle different perspectives of the market.

Cryptocurrencies are a new "invention" and every new innovation brings a specific set of risks with it. Chapters 3 and 4 give a broad overview of the different qualitative and quantitative risks in cryptocurrency land. Everyone who is active in the market knows that in the past cryptocurrencies have exhibited very volatile periods. There have been flash crashes, even pump-and-dumps are not uncommon and extreme fluctuations are much more frequently occurring than in other asset classes. Hence we are dealing with a variety of much more pronounced market risks.

---

[1] Satoshi (2009) Bitcoin: a peer-to-peer electronic cash system.

Chapter 5 investigates the behaviour of options and futures on Bitcoin. These are not only traded on the so-called cryptocurrency exchanges but also on one regular exchange, namely the Chicago Mercantile Exchange.

Investors are very interested in cryptocurrencies, as they might diversify a portfolio. On top of that, cryptocurrencies could enlarge the profit of a portfolio possible at the cost of a more volatile portfolio. Therefore, Chap. 6 extensively compares different portfolios construction mechanisms with the usual asset classes and Bitcoin.

The last chapter introduces the reader to further related literature.

Brussel, Belgium   Eline Van der Auwera
Heverlee, Belgium   Wim Schoutens
Ispra, Italy   Marco Petracco Giudici
Ispra, Italy   Lucia Alessi
April 2020

Contents

Acronyms

ACF	Autocorrelation function
ADF	Augmented Dickey–Fuller
AF	Autocorrelation function
AIC	Akaike information criterion
AML	Anti-money laundering
APE	Average pricing error
AR	Autocorrelation
ARMA	Autoregressive moving-average
ASIC	Application specific integrated circuit chips
BFT	Byzantine fault tolerance
BIC	Bayesian information criterion
BIS	Bank for International Settlements
BRR	Bitcoin reference rate
BTC	Bitcoin
CBOE	Chicago Board Options Exchange
CME	Chicago Mercantile Exchange
CPU	Central processing unit
DAO	Decentralized autonomous organizations
dAPPS	Decentralized applications
DPoS	Delegated proof-of-stake
ESMA	European Securities and Markets Authority
ETC	Ether classic
ETF	Exchange traded fund
ETH	Ether
FBA	Federated Byzantine Agreement
GARCH	Generalized autoregressive conditional heteroskedasticity
GSADF	Generalized supremum augmented Dickey–Fuller
HQIC	Hannan–Quinn information criterion
ICO	Initial coin offering
IPO	Initial public offering
KS	Kolmogorov–Smirnov

KYC	Know your customer
LTC	Litecoin
PACF	Partial autocorrelation function
PoS	Proof-of-stake
PoW	Proof-of-work
RMG	Royal mint gold
RMSE	Root mean squared error
SPV	Simple payment verification
VaR	Variance-at-risk
VAR	Vector AutoRegression
USC	Utility settlement coin
USDT	Tether
XLM	Lumens (Stellar)
XMR	Monero
XRP	Ripple

Part I
Introduction to Cryptocurrencies

Chapter 1
Blockchain

Abstract In 2008, the world was introduced to a new concept: Bitcoin. Although it took several years to become known among a broad public, it is considered to be a revolutionary idea. This section gives a broad overview of the Blockchain technology behind Bitcoin.

Keywords Blockchain · Hash pointers · Hash algorithm · Distributed ledger · Consensus protocol · Proof of work · Proof of stake · Byzantine fault tolerance · Delegated proof of stake

In 2008, the world was introduced to a new concept: Bitcoin (Satoshi 2008). Although it took several years to become known among a broad public, it is considered to be a revolutionary idea. Actually, Satoshi (2008), the proclaimed inventor of Bitcoin, never intended to create a digital currency, he merely wanted to develop a peer-to-peer electronic cash system.

Cryptocurrencies are a side-product of digital cash. In the nineties, many people have tried but failed to create a decentralized digital cash system (BlockGeeks 2019). The newly introduced blockchain technology makes this decentralisation now possible; the fact that information is distributed and not copied, is exactly what makes blockchain technology so revolutionary. Regular currencies, like Euro or Dollar, only work because people trust the currency and double spending is annihilated through a central authority. This trust is ensured by governments and laws. Cryptocurrencies actually do not need such trust anymore, they rely on cryptographic proof. Moreover, cryptocurrencies promise its users to some extent anonymity. In addition, no government interference is needed anymore.

Cryptocurrencies are mathematically protected digital currencies that are maintained by a network of peers. Digital signatures authorize individual transactions and ownership is passed via transaction chains. The ordering of the transactions is protected in the blockchain. By requiring difficult mathematical problems to be solved within each block, the attackers are racing against the rest of the network to solve computationally difficult problems, they are unlikely to win.

1.1 Blockchain in a Nutshell

Blockchain is a way of recording information on lots of devices, all at once, through the internet. The basic process of blockchain is as follows (Lisk 2018). Suppose someone makes a transaction, this transaction must then be included on the blockchain. A transaction can include a cryptocurrency exchange, contracts, records or any kind of transfer of information. The requested transaction is then broadcasted to a worldwide peer-to-peer network consisting of computers, also known as nodes. These nodes have to verify the transaction as legitimate as well as the user's status by means of known algorithms. All the nodes perform the same activities and they all store a copy of the ledger. Once consensus between the nodes is reached, it becomes part of a block of data which contains other transactions on the ledger. If the block is complete, it competes with other blocks to become the next block of the existing blockchain. Once the block is attached and secured using cryptography, it is permanent and unalterable.

A blockchain can be viewed as a spreadsheet duplicated thousands of times across a worldwide network of computers. This spreadsheet is regularly updated, such that new transactions can become part of the spreadsheet. The decentralized part of blockchain is a consequence of the fact that the system constitutes of a worldwide peer-to-peer network and therefore multiple locations store the same information. Thus, it ensures that the information is easily verifiable and public. As a result, the information on the blockchain is hard to corrupt. It would need a massive amount of computer power to alter the recordings. The name blockchain actually comes from the way data is stored, namely blocks hold information and are thoroughly linked by chains.

Each block in the blockchain contains transactions, a time stamp, a digital signature to identify the account who did the recording and a unique identifying link. This link, usually created by hashing, will point to the previous block in the chain and ensures that information is unalterable. Therefore, blocks further down the chain are more secured than more recent blocks because many other blocks point to it.

1.2 Network and Nodes

Blockchain works peer-to-peer, there is no single central authority within the network. Information is constantly recorded and interchanged between all the participants. The users of the network are the backbone of the whole system, they trade a part of their computing resources to keep the network running. In some cases participants have the opportunity to collect transaction fees or rewards in exchange for the computing power. Nodes can have different roles in the network (Lisk 2018). Full nodes, or miners, are permanently connected and store the entire blockchain, they verify and propagate the activities and blocks in the network (S 2018). Simple

Fig. 1.1 Different types of graphs. (**a**) Decentralized. (**b**) Centralized. (**c**) Distributed ledger

payment verification (SPV) nodes, on the other hand, do not store the entire blockchain. Therefore, they rely on full nodes to receive and propagate transactions throughout the network. SPV nodes basically download the transactions which are significant for them (Satoshi 2008). All nodes are considered equal although some have different tasks. The basic tasks of a node are:

- Check the validity of transactions and add them to existing blocks or simply reject them.
- Save and store blocks of transactions.
- Broadcast and spread the transaction history to other nodes to secure synchronicity.

A user who is interested in full autonomy and authority should run a full node. Note that depending on the consensus algorithm, the requirements to own a node may be different.

Nodes form a random graph in the sense that a random node is connected to other random nodes (Javarone and Wright 2018), like Fig. 1.1c. All nodes are interconnected in order to verify the transactions and receive them, unlike in Fig. 1.1a, b.

Nodes are able to come and go to the network in order to allow for a flexible and dynamic system. An active (online) node which goes offline, is forgotten after a predetermined amount of time to ensure a smooth working system. On the other hand, when an offline node comes back, it needs to get back up to speed. The node has to download all the blocks that were added to the blockchain while the node was offline. Theoretically, a single node can keep the blockchain afloat, however, the network would then be highly vulnerable to corruption.

A new transaction is propagated through the network by moving over all the connections between the nodes. Eventually every node is connected to all the nodes in the network by using its peers. Once a node receives a transaction, it checks the senders righteousness and if the money has not yet been spent. Afterwards, the node sends the information to its peers until the whole network knows about the transaction. The time it takes for a whole block of transaction to go through the

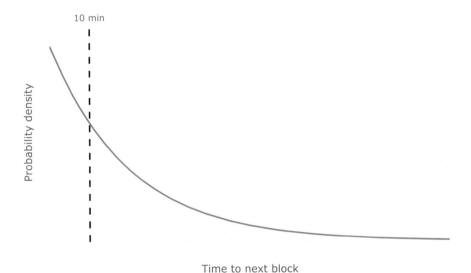

Fig. 1.2 Probability density function of the time until the next block is mined in Bitcoin. The dotted line represents the 10 min time point

whole network can be quite long, this algorithm is called the flooding algorithm. For example, for Bitcoin the probability density function is shown in Fig. 1.2 (Narayanan et al. 2016). Double spending is avoided because only the transaction which links first to a certain unspent transaction, is processed. A node will also reject an unusual script. Moreover, nodes, who act dishonest, will be punished. An example of a punishment is reclaiming the award for blocks received in the past (Drescher 2017).

A new block needs to follow the same rules to get its place in the blockchain as transactions. However, it also needs to obey a consensus criterion, which validates if all the transactions in the block are righteous and if the block is built on the current longest chain.

1.3 Cryptographic Algorithm

Cryptography is the method of disguising and revealing information by using mathematics (Lisk 2018). In blockchain technology, cryptography is used with a dual purpose. First, it ensures that the identity of the practitioners is hidden and second it assures that information is secured as well as the transactions.

The mostly used cryptography method, in case of cryptocurrencies and blockchain, is public-key cryptography (Lisk 2018). This method allows information to be passed alongside with the public key while the private key stays with the sender. Two different keys encrypt and decrypt the information, the

Fig. 1.3 Creation of the hashed digest with the digital signature placed on it (Lisk 2019)

sender's public and private key encrypt the information, while the private key of the receiver together with the sender his public key are used for decryption. Every transaction initiates the key generation algorithm which creates a new public and private key for the sender. Moreover, the public-private-key encryption algorithm places a digital signature on a hashed digest to ensure its integrity, the hashed digest is a product of the hashing algorithm performed on the data, see Fig. 1.3. The document itself, together with the private key of the user, is used to construct the signature. Hence, the signature will not match anymore if the data is altered. Moreover, the receiver of the data is able to verify the authenticity of the document by analyzing the digital signature and the data. The receiver uses the public key to decrypt the digital signature and the hash algorithm to scan the document. If both outcomes match, then the receiver knows that the content of the message is not corrupted in transit.

1.3.1 Hash Algorithm

The hashing algorithm takes an input of arbitrary length and transforms this by using mathematical transformations to an output of a fixed length. Bitcoin, in particular, uses SHA-256 hashing algorithm. This particular algorithm translates every transaction to an output of seemingly random numbers of 256 bits. The input, in general, can be a transaction or a document but it can even be a block of transactions. The algorithm is in some sense deterministic, because the same input will always generate the same output. It is almost impossible to determine the actual length of the input due to the fixed length of the hashed data. A good hashing function has the following five main properties (Su 2017):

- **Deterministic**: Changing one single detail in the input completely changes the hashed output. Hence, the hashing function provides some kind of security. Suppose that the output would only be changed a little bit if one changed the input then the input would be much easier to retrieve.
- **Computationally efficient**: The time elapsed to produce the output should be small.
- **Collision resistance**: A collision happens when a hash function maps two different inputs to the same outcome. This is extremely hard theoretically and in practice almost impossible to find two different inputs in the hash algorithm to

give the same output. The phenomenon can still occur due to the fixed length of
the output.

- **Hiding feature**: It is impossible to find the input value based on the hash value.
  Theoretically it is possible, in practice however this is computationally infeasible.
- **Random looking output**: It is hard to predict the output of the algorithm.

In the particular case of cryptocurrencies, cryptographic hash functions are used
(Berlanger and Kennis 2016). One of the cryptographic hashing algorithms main
objectives is to let the receiver check the authenticity of the information. In other
words, this algorithm is implemented for information security, authenticity control
and other security measures. The user checks the information by running it through
a hashing algorithm. If the product of this algorithm matches the hash output send
by the sender, the receiver is sure that the information has not been tampered
with throughout the process. A cryptographic hash function $H(\cdot)$ has the following
properties:

- **Pre-image resistant (one-way)**: A hash function H is said to be pre-image
  resistant if for a given h in the output space, it is hard to find an x such that
  $H(x) = h$.
- **Second pre-image resistance**: A hashed function H is second pre-image
  resistance if for a given input x_1, it is hard to find another input $x_2 \neq x_1$ such that
  $H(x_1) = H(x_2)$.
- **Collision resistant**: A hash function H is said to be collision resistant if it is
  infeasible to find two values, x_1 and x_2, such that $x_1 \neq x_2$ but $H(x_1) = H(x_2)$.

In other words:

- **Pre-image resistance (one-way)**: Given the hashed output, it is impossible to
  find any information regarding the original message.
- **Second pre-image resistance**: Given a certain input, it is extremely difficult to
  find another input to provide the same hashed output.
- **Collision resistance**: It is hard to find two different inputs which provide the
  same hashed output. Such a pair of inputs that do give the same output are called
  a cryptographic hash collision.

The system does not hash an entire block of the chain at once, it rather
hashes each transaction on its own. Afterwards each hashed transactions is linked
together, by using hash pointers, with the remaining hashed transactions in the
block. Blockchain most often uses a Merkle tree (see Fig. 1.4), also called a hash
tree, to represent the hashed and recorded data on the Blockchain. This concept
was introduced by Ralph Merkle in 1979. Merkle trees allow the user to validate
individual transactions without having to download the entire blockchain.

1.3.1.1 Hash Pointer

A hash pointer actually serves as a reference where data is stored. It has a dual
purpose, firstly it serves as some kind of information store and secondly it is able to

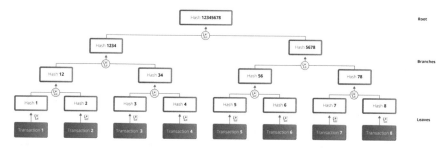

Fig. 1.4 Merkle tree taken from Lisk (2019)

verify that information has not changed. One of requirements for a properly working hash pointer is that the chain has no cycles. In a blockchain, hash pointers help to secure the chain, since a blockchain is a sequence of blocks where the hash pointers link several blocks together and contains a digest of that value. Particularly, the previous block is hashed together with a new block in the chain and this happens for every two sequential blocks in the chain. As mentioned before, hashing makes the chain more secure because tampering with one block in the chain would mean that all the later blocks also needs to be changed. The first block in the chain, the genesis block, can be assumed to be not tampered with.

In Merkle trees each block has two different hash pointers to ensure the structure of the tree. A Merkle tree as in Fig. 1.4 thus encodes the data more securely and efficiently.

As mentioned before, the first block of a blockchain is called the genesis block. This block contains transactions which form a unique hash. This hash forms together with all new transactions the input for another hashing. In other words, in a blockchain each hash has a link to the previous block. This basically explains the name blockchain: every new block is linked through the hash to the previous block.

1.4   Distributed Ledger

The distributed ledger in any cryptocurrency does not contain a ledger of balances. The ledger actually stores all the transactions. Meaning that if someone wants to prove they own a certain amount of digital currency, one has to go through the ledger and show that that person has still enough unspent amount of the digital coin.

Suppose that the whole amount is not used in a transaction, in this case an additional output is created in the transaction, which transfers the change back to the sender. Note that a transaction is distributed to the whole network because the nodes need to verify if the transaction is honest. In other words, the ledger contains the total history of transactions on the blockchain.

1.5   Consensus Algorithms

Consensus algorithms ensure that all the nodes in the network are synchronized. Meaning that each node has to option to include or exclude a new transaction to their copy of the ledger. When the majority of actors chose to include the transaction, consensus is achieved and the transaction is added to every copy of the ledger. The reached consensus can be seen as the truth (Lisk 2018). In order to execute a consensus protocol, several rules for communication and transfer of data between electronic devices must be obeyed. This system actually provides that the blockchain is 'self-auditing', which allows the blockchain to be updated, while assuring the honesty of the chain. Another advantage of the consensus protocol is that blockchains are extremely hard to corrupt.

In general, a consensus protocol can only have two of the following properties: safety, fault tolerance or liveness. Fault tolerance means that the protocol remains operable if a validator fails. Safety refers to the exclusion of bad experiences to the blockchain, like a fork in Fig. 1.8. For that reason, assuming that the protocol is safe, the validators can not create two different ledgers. In case a fork happens, the network will stop making progress. Liveness, on the other hand, is the guarantee that ledgers are closed. Validators can diverge and create an accidental fork while the network keeps working. Hence, liveness and safety cannot be both characteristics of a consensus protocol.

One important aspect is the reward attached to transactions to keep the network secure. The incentives exist to avoid invalid confirmations, like *double spending*. Double spending occurs when a certain cryptocoin is spend twice (Investopedia 2019). It is a problem unique to digital currencies since physical money is harder to replicate and easier to determine its righteousness. In general, consensus algorithms are either difficult to replicate or extremely costly to carry out. There exist many different consensus protocols, in the next sections the most common ones are explained.

1.5.1   Proof-of-Work

Proof-of-Work (PoW) is the consensus algorithm implemented in Bitcoin. It was introduced by Bitcoin.com (2017). Today, this algorithm is considered by some problematic due to its enormous energy drainage. Many recent cryptocurrencies do not use this consensus algorithm anymore; Ethereum is moving towards another algorithm. Proof-of-Work chooses the liveness and fault tolerance characteristics. In 2017 alone, Bitcoin has experienced 11 forks, like the fork resulting in Bitcoin and Bitcoin Cash. The liveness property is the reason why a transaction is truly confirmed after adding several blocks.

In the PoW algorithm, the process is called mining and the nodes are called miners. Miners solve very hard mathematical puzzles in order to create a new

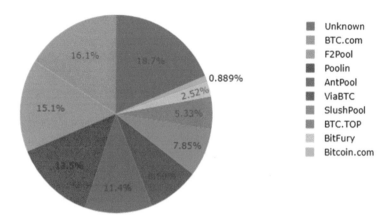

Fig. 1.5 The hashrate distribution in Bitcoin mining

block. In other words, blocks are very hard to produce. As an incentive to trade computational power, miners receive a reward when they mine a new block. The reward can be predetermined, as is the case in Bitcoin, or it can be equal to the transaction costs of all the transactions in the block. The correctness of the blocks is easy to verify. The answer to the mathematical puzzles is found by trial and error, hence, it is extremely energy draining and uses lots of computational power. In the PoW consensus algorithm the longest chain wins, which means that miners continue to add blocks on top of the longest chain in existence. This chain is considered to be the most trustworthy because every block that wants to be added to the chain needs to be verified and the verification process is easy to perform.

As mentioned, PoW has its issues. The most prominent problem is that PoW is not efficient, it needs lots of energy and computer power (Blockgeeks 2019). It has been reported that a single Bitcoin transaction can cost as must energy as a Dutch household consumes in 2 weeks (Lisk 2018). Due to the high energy need, participants want a substantial amount of coins as reward to keep participating. If the energy consumption would go down, the reward can also go down. Another problem is that some companies and people mine blocks easier and faster because they have better equipment. Therefore, some users have a clear advantage over others, in other words, the blockchain becomes more centralized. Figure 1.5 clearly shows that on 10 December 2018 only 3 mining pools mine over 50% of the blocks in Bitcoin according to Blochchain.info. Moreover, the specialized equipment is built with the sole purpose of mining. Hence, if the tools become outdated, they are simply useless and become electronic waste.

In order to let your transaction valid faster you can opt to add a transaction fee which goes to the miner who can validate the block first.

During the boom in cryptocurrencies in 2017, the transaction fees in Bitcoin became roughly 25 dollars, see Fig. 1.6 and the transactions also experienced significant delays as shown in Fig. 1.7 (Rauchs et al. 2018). Ethereum experienced

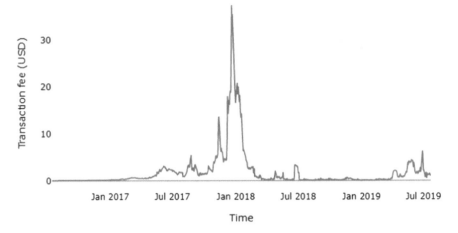

Fig. 1.6   Transaction fee in Bitcoin

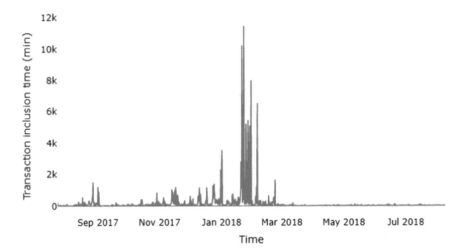

Fig. 1.7   Transaction inclusion time of Bitcoin

problems of its own during this boom, a particular gaming application became too popular, which resulted in a clog of the blockchain for nearly 2 days.

1.5.2   Proof-of-Stake

The Proof-of-Stake (PoS) algorithm was first used in 2011 in a coin called Peercoin. This algorithm is beneficial over Proof-of-Work in the sense that it is more energy efficient, secure and reduces centralization. The main differences with the PoW algorithm are that miners are replaced with validators and the next validator of a block is chosen in a deterministic way. The system is called Proof-of-Stake because the more currency a user owns, the higher the amount at stake if the network is under attack (Martinez 2018).

In order to become a validator in the PoS algorithm, the user needs to have a certain amount of the currency at stake (Blockgeeks 2019). Afterwards, the validators start to check all the presented blocks and try to find which block needs to be added to the chain by placing bets on the blocks. The probability of receiving the reward is proportionate to the amount of currency the validators put at stake. For example, if you have three validators, where the first validator has 3 coins, the second validator inserts 2 coins and the third has 5. Then, the probability that the first validator is chosen to validate the block is 30%, the second has 20% chance and the third 50%. In other words, some individuals have an higher chance to be chosen than others. However, everyone has a chance at the reward which is basically a part of all of the transaction costs and due to this randomization the PoS algorithm prevents centralization. This system does not require many new coins to come into existence to keep validators incentivized. Therefore, the price of a coin can be more stable (Lisk 2018).

If an individual wants to perform a 51% attack in this system, he needs to own the majority of all the coin in existence and this would be very costly to achieve. Moreover, the other individuals would either exit the currency, if they notice, or will drive up the price to prevent it. In general, the user with the highest amount of currency, has the highest incentive to keep the network secure. Furthermore, it is also possible to use economic penalties to make the attack more expensive.

The PoS consensus protocol has one major flaw, namely the "nothing at stake" problem (Blockgeeks 2019; Martinez 2018). This event occurs when the consensus fails. The block generators in this case are indifferent between supporting various blockchain histories. In other words, many validators build on every fork if forks occur, see Fig. 1.8. This problem does not occur in PoW, because it takes lots of energy to build new blocks. Hence, building on a rejected chain would be wasted energy and as a result no one is willing to do it. In PoS, on the contrary, it costs nothing to validate a block. Moreover, building on every fork is financially the most interesting. If validators mine on several forks at once, they increase the chance of securing transaction fees from the winning fork. The nothing at stake problem delays and complicates reaching consensus. In order to decrease the probability that

Fig. 1.8 Representation of a forked blockchain

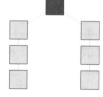

all validators built on all forks, validators lose a portion of their currency if they bet on every fork.

1.5.3 Byzantine Fault Tolerance

The Byzantine fault tolerance consensus protocol is a solution to the Byzantine generals problem stated by Lamport, Shostak and Pease in 1982. The problem is as follows (Konstantopoulos 2017): there is an army consisting of one general and $n-1$ lieutenants scattered around the city. The army is planning to attack a common enemy and it still has to decide when the attack will take place. The attack is only successful if the whole army charges synchronously. The general and his lieutenants need to reach a consensus on the best time to attack by sending messages back and forth. However, some of the lieutenants are traitors, meaning that they can lie about their choice. The Byzantine generals problem replicates the problem in blockchain, meaning that a network (the general and his lieutenants) needs to reach a consensus on a broadcasted transaction (the time to attack), while some nodes can be faulty (traitorous lieutenants). Byzantine fault tolerance is the characteristic which defines a system that tolerates a certain number of failures who belong to the Byzantine generals problem.

The system is able to reach consensus as long as the number of dishonest nodes is small. A certain number of nodes needs to decide whether to include a transaction to the blockchain or not. In Byzantine fault tolerance (BFT), messages are send back and forth between nodes. Nodes use voting power to reach consensus. A new transaction is only added to the chain if over $2/3$ of all the validators agree. Consensus is reached much faster in BFT than in PoW because sending messages enables to reach consensus. One major problem of BFT is that it is more centralized. Moreover, validators must be part of a recommended validator list, which means that new validators are selected.

Federated Byzantine agreement (FBA) (Lumenauts 2018) is a modified form of Byzantine fault tolerance, which is the consensus protocol in Stellar and Ripple. Unlike BFT, FBA protocol is decentralized, all nodes can become validators once they are added to an existing quorum slice. Meaning that each node, on its own, can decide which other nodes to trust. The list of trusted validators is then called a "quorum slice". The quorum slice of validators overlap most of the time. This is how a network wide consensus on a transaction, without the need of a centralized authority, is formed.

Nodes do not have to be known upfront and membership can be open (Shaan 2018). In Ripple, the nodes are preselected by the Ripple foundation, while Stellar leaves membership open. FBA and BFT favour safety over liveness (Lumenauts 2018). In case of an accidental fork, the progress in the network halts, until consensus can be reached. Safety also ensures that you do not have to wait for a transaction to be confirmed. In other words, once consensus is reached, the transaction is confirmed. Safety shortens the transaction time enormously.

In BFT or FBA, an attack using computing power is not possible, because validators sign new transactions using their private keys and cryptographic puzzles do not need to be solved. Dishonest validators are not able to retrieve the private keys of over 2/3 of the total number of validators. As a result, BFT and FBA are called asymptotically secure.

1.5.4 Delegated Proof-of-Stake

Delegated Proof-of-Stake (DPoS) is a proposed solution to the scalability problem in distributed ledgers. It increases the speed of finalizing transactions substantially. However, the scalability comes at a cost, the network becomes more centralized than Bitcoin, for example. Delegated Proof-of-Stake has been proposed by Dan Larimer in 2013 (Samani 2018). DPoS chooses for safety and fault tolerance, accordingly, users can choose to fork if they disagree with the majority.

Delegated Proof-of-Stake combines real-time voting and a system of reputation to reach consensus (Konstantopoulos 2018). Every token owner in this protocol has a certain amount of power in the block producing process. The voting power of a token owner is proportionate to the number of tokens he or she possesses. Moreover, voters can give a proxy to another voter, to vote on their behalf. Block producers, also called witnesses, are selected by the token owners and are responsible for creating and signing new blocks. The number of block producers is limited, as a consequence this system is not fully decentralized. Furthermore, witnesses have to prove they have the best interest of the network at heart (Lisk 2018). Block producers can be voted in or out at any time. DPoS uses the threat of losses and reputation damage as incentives to keep the system honest.

Block producers do not have the power to exclude a certain transaction from the blockchain, they can only delay its inclusion. The number of block producers depends on the specific cryptocurrency. In EOS (see Chap. 2), for example, there are 21 witnesses, while Lisk, another cryptocurrency, has 101 block producers. A block is finalized if 2/3 of the block producers plus one vote on it (Konstantopoulos 2018). The block producers take turns in producing a block at predefined time slots, this is how the longest chain is created. Suppose a fork occurs, than the other block producers still build further on the longer chain (the honest chain).

DPoS is not the perfect system, it exchanges decentralisation for scalability. Therefore, the system is easier to corrupt than Proof-of-Work. For example, if many users of the system are bribed, than the majority of block producers can be dishonest. Moreover, there can be an unlimited amount of forks. However, this behaviour will not last long, since these corrupt block producers will be voted out.

1.5.5 Summarizing Table

Table 1.1 Summarizing table of the consensus algorithms

Algorithm	Nodes	Next block "creation"	Incentive	Cons	Pros
Proof of work (PoW)	Miners	Solve computationally intensive puzzle	Reward	Energy draining, costly, 51% attack	Most decentralized
Proof of stake (PoS)	Validators (chosen deterministic)	Validation	Coins at stake, transaction costs	Nothing at stake problem, rich players control consensus	Less energy draining, faster than PoW, less hardware is needed
Byzantine fault tolerance (BFT)	Approved validators	2/3 of validators agree	No reward, sometimes small amount of currency is destroyed	More centralized than PoW	Low cost, scalable, less energy draining, fast confirmation
Delegated proof of stake (DPoS)	Witnesses (limited number)	Vote on producer who generates a block by validating transactions	Reward	More centralized than PoW, easier to perform 51% attack than PoS, cartel formation	Benefits of PoS + secure real-time voting and rewards are evenly distributed

References

Berlanger N, Kennis, I-M (2016) The blockchain technology. Master's thesis, KULeuven

Bitcoin.com (2017) What is bitcoin? https://www.bitcoin.com/info/what-is-bitcoin

Blockgeeks (2019) Basic primer: Blockchain consensus protocol. https://blockgeeks.com/guides/blockchain-consensus/1-Proof-Of-Work

BlockGeeks (2019) What is cryptocurrency: everything you must need to know! https://blockgeeks.com/guides/what-is-cryptocurrency/

Drescher D (2017) Blockchain basics: a non-technical introduction in 25 steps. Apress, New York

Investopedia (2019) Double spending definition

Javarone MA, Wright CS (2018) From Bitcoin to Bitcoin cash: a network analysis. In: Proceedings of the first workshop on cryptocurrencies and blockchains for distributed systems. Association for Computing Machinery, New York, pp 77–81

Konstantopoulos G (2017) Understanding blockchain fundamentals, part 1: Byzantine fault tolerance

Konstantopoulos G (2018) Understanding blockchain fundamentals: delegated proof-of-stake. https://medium.com/loom-network/understanding-blockchain-fundamentals-part-3-delegated-proof-of-stake-b385a6b92ef

Lisk (2018) https://lisk.io/academy/blockchain-basics/how-does-blockchain-work/delegated-proof-of-stake

Lisk (2018) Blockchain cryptography explained. https://lisk.io/academy/blockchain-basics/how-does-blockchain-work/blockchain-cryptography-explained

Lisk (2018) Proof-of-work. https://lisk.io/academy/blockchain-basics/how-does-blockchain-work/proof-of-work

Lisk (2019) www.lisk.io

Lisk (2018) Blockchain basics. https://lisk.io/academy/blockchain-basics

Lumenauts (2018) How the stellar protocol works. https://www.youtube.com/watch?v=X3Gj2nQZCNM

Martinez J (2018) Understanding proof of stake: the nothing at stake theory. https://medium.com/coinmonks/understanding-proof-of-stake-the-nothing-at-stake-theory-1f0d71bc027

Narayanan A, Bonneau J, Felten E, Miller A, Goldfeder S (2016) Bitcoin and cryptocurrency technologies: a comprehensive introduction. Princeton University Press, Princeton

Rauchs M, Blandin A, Klein K, Pieters DG, Recanatini M, Zhang B (2018) Second global cryptoasset benchmarking study. Technical report, University of Cambridge

S J (2018) Blockchain: what are nodes and master nodes? https://medium.com/coinmonks/blockchain-what-is-a-node-or-masternode-and-what-does-it-do-4d9a4200938f

Samani K (2018) Delegated proof of stake: features and tradeoffs. https://multicoin.capital/2018/03/02/delegated-proof-stake-features-tradeoffs/

Satoshi N (2018) Bitcoin: a peer-to-peer electronic cash system. Technical report, bitcoin.org

Shaan R (2018) Federated Byzantine agreement. https://towardsdatascience.com/federated-byzantine-agreement-24ec57bf36e0

Su HL (2017) What is a cryptographic hash? http://homeowmorphism.com/articles/17/What-Is-A-Cryptographic-Hash

Chapter 2
Types of Cryptocurrencies

Abstract The most well-known cryptocurrency is Bitcoin. This coin has the first mover advantage over all the other coins which were created afterwards. Today there exists a huge variety of digital coins. Some of them have found their way in the digital world, others hardly have any value at all or are scams. This chapter first provides general information on cryptocurrencies with a focus on those with the highest market value. It will detail users, objectives and opportunities of cryptocurrencies. Moreover, it will explain how new cryptocurrencies can be created. Thereafter, this chapter will give an overview of the characteristics of different cryptocurrencies. The chapter concludes with an overview of the different attributes of the mentioned cryptocurrencies.

Keywords Irreversibility · Pseudononymous · Permissionless · Commodity money · Means of payment · Security · Fork · ICO · Users of cryptocurrencies · Objectives of cryptocurrencies · Opportunities · Bitcoin · Ripple · Litecoin · Ethereum · Monero · Stellar · EOS · Tether

The most well-known cryptocurrency of the year 2019 is Bitcoin. This coin has the first mover advantage over all the coins which are created afterwards. Today there exists a huge variety of digital coins. However, many of them hardly have any value at all. This chapter first provides general information on cryptocurrencies. It will detail users, objectives and opportunities of cryptocurrencies. Moreover, it will explain how new cryptocurrencies can be created. Thereafter, this chapter will give an overview of the characteristics of different cryptocurrencies. The cryptocurrencies selected in this chapter have the highest market value in September 2019. The chapter concludes with an overview of the different attributes of the mentioned cryptocurrencies. Besides the below mentioned coins, there also exist projects related to corporate coins. A popular example of these corporate coins is Libra, a coin designed by Facebook among others. We will, however, not elaborate on this subject because these projects have a lot of uncertainty on a regulatory, conceptual and market point of view.

E. Van der Auwera et al., *Financial Risk Management for Cryptocurrencies*,
SpringerBriefs in Finance, https://doi.org/10.1007/978-3-030-51093-0_2

2.1 General Information on Cryptocurrencies

Cryptocurrencies are called CRYPTOcurrencies because the consensus algorithm is backed by extreme forms of cryptography. Digital coins are not secured by humans or central banks but by mathematics. Every digital coin shares a common set of characteristics. There are transactional properties and monetary properties (Cryptocurrency Army 2019; BlockGeeks 2019), as shown below.

The transactional characteristics are:

- Irreversibility: once a transaction is confirmed, it is final and cannot be altered anymore.
- Pseudonymous: transactions and accounts are not linked to real world names and addresses.
- Fast and Global
- Secure: funds are locked in a public key cryptography system. Only the owner of the private key can send the cryptocurrencies.
- Permissionless: everyone can download the software for free and use it. However, some cryptocurrencies are not truly permissionless, in the sense that their ledger is maintained by a network of authorized stakeholders.

While the monetary properties are:

- Controlled supply: most cryptocurrencies limit the amount of cryptocurrencies available in total, while other cryptocurrencies cap the yearly amount of cryptocurrencies to be mined.
- No debt but bearer: Fiat money on a bank account represent a debt by the bank to the owner. Cryptocurrencies do not represent debt, they represent themselves. It is hard to characterize cryptocurrencies, there is a discussion in the literature whether cryptocurrencies have a store of value, if they can be considered as a means of payment or if they are a security or commodity. We will elaborate more on this in Sect. 3.2.

Moreover, cryptocurrencies are divided into three categories, namely as a means of payment, a commodity or a security. Commodity money has another use than that of being merely a medium of exchange and it is naturally a scarce good. Fiat money is only useful as a media of exchange and it is scarce by design. Which means that an endless supply would only drive the value of fiat money to zero. Cryptocurrencies can be treated as a security because transactions on the blockchain can be proof of ownership.

Cryptocurrencies can be considered as a means of payment if it is a unit of account, store of value or a medium of exchange. However, there is only a small number of retail shops which do actually accept cryptocurrencies. According to the ECB (Mersch 2018) none of the cryptocurrencies pass the "unit of account test", retailers would take on significant risks if they truly start accepting cryptocurrencies since cryptocoins are not backed by any central bank. The store of value part is perhaps the most important part which cryptocurrencies do not adhere. It is difficult to pinpoint what the intrinsic or extrinsic value is of cryptocurrencies.

The first strand in the literature claims that cryptocurrencies can be regarded as a means of payment, hence it must have a store of value. Barber et al. (2012) support this conclusion, they state that the core of cryptocurrencies and in particular Bitcoin could support a robust decentralized currency but the parameters embedded in the system are poor and could use some improvement. Grinberg (2011), on the other hand, states that Bitcoin is not likely to replace e-commerce services like PayPal, because most customers do not care about centralisation or anonymity. Real customers do not want to shop goods in a currency they do not fully know. However, he also believes that Bitcoin can become a real competitor in the micro-payment and the virtual/game-related market.

Selgin (2015) asserts that cryptocurrencies are both a commodity and a currency. He introduces "synthetic commodity money", this type of money involves features of both fiat and commodity money without completely fitting in one category. Synthetic commodity money would be naturally scarce and can only be used as a medium of exchange. Selgin claims that cryptocurrencies are an example of semi-inelastic synthetic commodity money since the technology behind it makes it scarce, it can be used as a means of payment and the rate of creation declines to zero (in most cases). However, from a macroeconomic perspective cryptocurrencies are not the perfect monetary medium because the demand is unpredictable and therefore the price heavily fluctuates.

Glaser et al. (2014), on the other hand, find that uninformed parties (and new users) mostly use cryptocurrencies as an alternative investment option and thus not as a payment medium in 2014. The high exchange rate volatility of cryptocurrencies (and more in particularly in Bitcoin) is a first indication that cryptocurrencies are used as a speculative asset. Another indication is the fact that Bitcoin (and other cryptocurrencies) react to news events (see Chap. 7).

2.1.1 Creation

New coins come into existence through an initial coin offering (ICO) (see also Sect. 3.3.3). A new cryptocurrency can also be created through a "fork". The new coin is then created from a previously established cryptocurrency. Basically, when a new rule or feature of a pre-existing currency is made, a fork occurs. There are two different kinds of forks, a hard fork and a soft fork. The soft fork is often a software update which does not lead to incompatibility issues. A hard fork, on the other hand, often directs to a new cryptocurrency which can either inherit the ledger of its predecessor or become a new cryptocurrency completely. Bitcoin and Bitcoin Cash are a popular example of the latter kind.

2.1.2   Objectives

Cryptocurrencies can differ in their objectives. However, it is possible to divide them into five categories based on the problems they tackle in their design (Elliott et al. 2018).

- **Digital cash coins**: The cryptocurrency is an alternative to cash, which does not rely on any government or central bank. These type of digital coins also offer a greater anonymity than the current payment system. Popular examples of this category are Monero and Bitcoin. These two coins will be compared and discussed extensively later in this section.
- **Payment infrastructure tokens**: These coins were developed to improve today's payment system. They offer payment transactions at either lower cost, higher speed or greater reliability or all three of these characteristics together. Most often, the conventional currency (such as a Dollar) is converted into the digital coin during the execution of payment. Examples of this type are Ripple and Utility Settlement Coin (USC). Ripple will be discussed in this chapter. USC is currently in development by a several banks to improve institutional payments.
- **Securities tokens**: A token of this category represents a unit of something of value. For example, a coin of Royal Mint Gold (RMG) represents the ownership of one gram of gold, held by the London Bullion Market Association.
- **Utility tokens**: These tokens provide the owner with future access to a product or service. The difference with the previous category lies in the future part. Securities tokens grand the user direct access to the good or service, utility tokens are often used to raise money for the development of a certain product. The tokens immediately give access to the product once it is finished. Examples are Filecoin and Golem. Filecoin and Golem plan to allow purchasing and selling of respectively data storage and computing power.
- **General platform tokens**: These types of cryptocurrencies support a platform which enables its users to create "apps". These platforms go beyond mere currency transactions, they automate more general transactions. For example, paying an insurance claim if certain conditions are met and signed. A popular example of this category is Ether (the token behind Ethereum).

The reason why tokens have value is because of supply, demand and trust. If no one wants to trade in a particular currency, its value drops until eventually zero. A coin is worth more granted that the demand is high and the supply is low. Moreover, trust in the coin's use and investors is also an important factor; a digital coin only has value if people believe it has value. Therefore, a cryptocurrency is, more than any other type of investment option, vulnerable to the effects of a confidence crisis.

2.1.3 Users of Cryptocurrencies

Cryptocurrencies have a broad reach, there are a lot of reasons for an actor to hold cryptocurrencies (Elliott and de Lima 2018). The two most obvious groups of buyers are speculator and investors. The first trade in cryptocurrencies to gamble on short-term gains, while the latter either uses them because they believe its price will rise, as an alternative asset, as a way to diversify their existing portfolio or to use as protection against other risks. However, cryptocurrencies also attract other people. The pseudonymity of cryptocurrencies for example attracts a different group of users. Bitcoins have been involved as in numerous illegal transactions. In general, this last mentioned group simply wants to be able to conduct transactions in anonymity.

In addition, some cannot buy all services or goods freely in their country, have certain capital restrictions, or are limited in investing their savings in foreign currencies as a save-heaven. In some troubled countries, cryptocurrencies could offer a way of securing ones wealth, because these governments are known to seize assets in times of crisis or political turmoil. In general, this group uses cryptocurrencies as an alternative currency.

Another group is represented by people who have a general distrust in the government. Cryptocurrencies are decentralized and therefore not subject to governmental interference. Some people believe that normal currencies give the government too much power and lead to systemic problems, like hyperinflation. Furthermore, cryptocurrencies are not subject to the mismanagement of governance, while still offering a store of wealth.

One last group are the technical users, for many blockchain applications one needs cryptocurrencies to pay for the service. Moreover, initial coin offerings also often require payments through cryptocurrencies. This groups incorporates those people who use cryptocurrencies for their technical benefits, like the efficiency in payments. The demand for cryptocurrencies is similar to the demand for definite commodities used in certain production processes. Each day more applications on blockchain come into existence, hence, the demand for cryptocurrencies in this group is also likely to rise.

2.1.4 Opportunities

One of the most obvious opportunities for cryptocurrencies lies in facilitating payment services. Cryptocurrencies can decrease the transaction costs and increase speed, transparency, reliability and immediacy in availability of funds. Especially, if one keeps in mind that many payments systems have been introduced many years ago. A certain amount of these systems are outdated and in need of more efficient procedures. Think about cross-border payments for example, these payments often travel through multiple banks in order to get to their destination. Cryptocurrencies are peer-to-peer and thus able to skip all those intermediate steps. This is much faster and in addition, it reduces part of the credit/counterpart risk involved in traditional payments.

However, one needs to keep in mind that transaction costs for some cryptocurrencies are not fixed. Transaction costs can even become substantial considering the price of the cryptocurrency. A first explanation can be found in to the number of transactions which need to happen at a certain time. In case the blocksize is limited (like for example in Bitcoin), it can take a considerable amount of time before a transaction is included and moreover, transactions with higher transaction fees are more likely to be included first. Other possible risks are the high energy usage and computational cost associated with Bitcoin or other coins. Hence rising electricity prices make it less attractive for miners to validate transactions with low transaction fees. One other disadvantage of some coins is that the settlement of transactions takes on average longer than certain regular payment settlements. Other coins try to overcome and circumvent these issues by experimenting with different consensus algorithms and using permissioned ledgers (non-public). If transaction costs are lowered and become less volatile, this could have promising opportunities for the financial world, for example in remittance transfers.

Another possible advantage of blockchain technology and cryptocurrencies in the financial world is in record-keeping of the ownership of stocks, bonds, etc. This would solve many outstanding problems related to the inability to keep track of "who owns what" in a timely manner (Kahan and Rock 2008). Moreover, debt securities and financial derivatives could be automatically executed using "smart contracts" on programmable blockchains. This would also make the ownership records more transparent and the trading itself less costly (Yermack 2015). Blockchain could also be used for real-time accounting which reduces the involvement of auditing companies (Yermack 2015).

There are also opportunities to be exploited out of the payment services world. Blockchain could be used for record management activities, because it is secure and highly reliable. For example for voting in elections, referendums or other government services. One could also store the digital identities of citizens on blockchains (Vitaris 2017). Moreover, the World Food Program, an organization of the United Nations, implemented a private Ethereum blockchain to provide food assistance (Das 2018). Using blockchain for humanitarian projects could save the organizations much money in the form of bank transfer fees. Another advantage is that the refugees could buy food using a bio-metric scan of their eye, because the data is stored on blockchain. Using this technique offers more security and privacy to the users.

2.2 Proof-of-Work Coins

2.2.1 Bitcoin (BTC)

Bitcoin (Bitcoin.com 2017; Vasu-Devan 2017) is probably the most well-known cryptocurrency. It was created in 2009 by an unknown individual or a group of individuals under the alias of Satoshi (2008). Bitcoin was the first digital currency and it can be used to purchase goods. Also many people regard the cryptocurrency as an investment good. Most other cryptocurrencies built further on the idea described in the white paper published by Satoshi (2008).

Bitcoin is a peer-to-peer electronic cash system, no central authority is needed. All the transactions are verified by its users through mathematics. Credit cards, for example, need a bank to verify its payments and most credit cards bear transaction costs. In Bitcoin, there is also transaction fee which is used as an incentive for the miners to keep the network secure, see Fig. 1.6. The faster the transaction must be confirmed the higher the fee. Credit card transactions can sometimes take quite some time to settle. The first confirmation of a Bitcoin transaction only takes about 10 min (in case enough transaction fee is paid and the network is not saturated). In other words, a new block is created in only 10 min. The creation of a new block rewards the miner also with a certain amount of Bitcoin, on the 10th of September 2019 this reward is 12.5 BTC. Experts believe the reward will be halved on 16 May 2020 (Half 2019). A rule of thumb to fully trust that a Bitcoin transaction is confirmed, is that it takes 6 blocks, hence approximately an hour. The block in which the transaction is recorded is then "deep" enough in the blockchain, such that it cannot be "reversed" anymore without a tremendous amount of energy and computer power. After only one block or a few blocks, although unlikely, this could still happen. This is the reason why Bitcoin transactions are in practice only finally accepted after a number of additional blocks (following the same chain) have been approved.

Bitcoin is a so called a deflatory currency. The reward of mining blocks is halved after a certain time and it will continue to do so until the reward becomes insignificant. Since only by mining new Bitcoins can be created the supply is limited. More precisely, its total amount is limited to 21 million Bitcoins. In September 2019, Bitcoin has an estimated market capitalization of 184.566 billion USD (Half 2019). Figure 2.1 represents the market capitalisation of Bitcoin over time.

Bitcoin is an highly volatile digital currency, it is not uncommon to see fluctuations of 20% on a single day as clearly seen in Fig. 2.2.

On the first of August 2017, Bitcoin experienced a fork, this fork led to the creation of Bitcoin Cash (Bitcoin.com 2019). Every user who held a certain amount of Bitcoin on the first of August 2017 received an equivalent amount of Bitcoin Cash on the forked blockchain and hence where at that point both holding the original Bitcoin and the new Bitcoin Cash. The main reason for this fork to occur is a discussion regarding the number of transactions possible a day. Bitcoin its blocksize limit is 1 MB a day, while Bitcoin Cash its blocksize limit is 8 megabyte a day. Therefore, Bitcoin Cash realized major benefits over regular Bitcoins, it offers faster

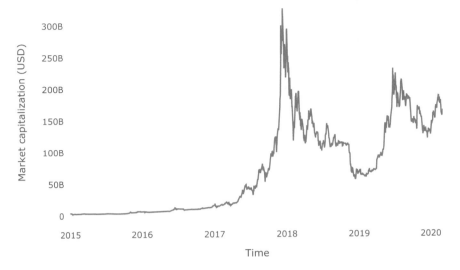

Fig. 2.1 Market capitalization of Bitcoin over time in USD

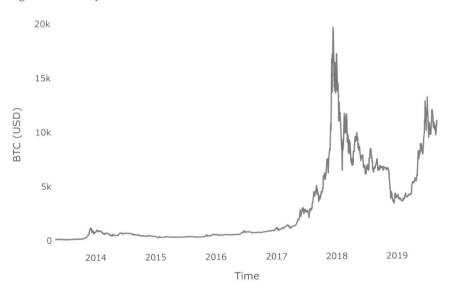

Fig. 2.2 Price evolution of one Bitcoin in USD (CoinMarketCap 2019)

payment settlements and lower fees. Interestingly, the value of the Bitcoin prior to this fork was equal to 2871.37 USD, whereas just after the value of a Bitcoin was 2737.76 USD and that of a Bitcoin Cash equaled 294.60 USD. Hence the holder of a Bitcoin before the fork was ending up with 2 coins with a total value of 3032.36 and actually profited 160.99 USD from the fork.

2.2.2 Ethereum

Ethereum Foundation (2018), Lewis (2015), and Rosic (2018) is founded by Vitalik Buterin on the 30th of July 2015. It is actually developed by a Swiss nonprofit organization called Ethereum Foundation. Ethereum is an open blockchain platform, this means that everyone can built and run applications, called dAPPs (decentralized applications), on this open source project. For this reason Ethereum is a general platform token.

As seen before, Bitcoin mainly stores transactions on the blockchain. Hence, in order to verify how much cryptocurrency someone owns you need to search for the unspent transactions. Ethereum, however, stores global states such as accounts, balances or codes on the blockchain. Like Bitcoin, it uses the Proof-of-Work as the consensus algorithm. However, a new block is created in only 14–15 s and it is smaller. Moreover, each new created block rewards the miner with 5 ETH, the cryptocurrency in Ethereum. Even blocks which are not adopted in the blockchain, called uncle blocks, give the miner a reward of 3–4 ETH. Additionally, referring to an uncle block in the main blockchain provides a reward of 0.15 ETH. The main reason for this reward is to encourage mining and decentralization of the blockchain. The main blockchain in Ethereum is the chain with the most work involved. A transaction in Ethereum costs on average 0.12 USD and is settled in approximately 6 min (O'Keeffe 2018).

Ethereum has its own cryptocurrency, namely Ether (ETH), it is mainly used as a mean of payment for the machines which execute small programs in the applications. In fact, there are two in-house currencies, namely Ether and Ether Classic (respectively denoted as ETH and ETC). Ethereum and Ethereum Classic used to be the same cryptocurrency. After a hard fork in the original version the community split in two (Rosic 2017). The developers of Ethereum performed a hard fork because a fraudulent user stole 50 million USD worth of Ethereum. One part of the community continued with the original chain where the 50 million was stolen, while the other part followed the updated version of the blockchain which started at the point before the hack. Moreover, there is no limit to the amount of Ether that can be created. However, the total amount of created Ether per year is limited to 18 million ETH. On 10 September 2019 Ethereum had a market capitalization of 19.645 billion USD (Coinmarketcap 2019).

Ethereum is a peer-to-peer blockchain where each node executes the same instructions. This massive parallelism makes the entire network slower and more computational intensive. However, it also makes the blockchain more secure against attacks and the blockchain has extreme levels of fault tolerance. Ethereum uses accounts and not addresses, like Bitcoin. To set up a contract, one needs to create an account. A dAPP is created by uploading the respective code on the account.

Every application does not run on a separate blockchain, they all run on the same platform. Bitcoin, on the contrary, is less flexible and adaptable since Bitcoin is not a programmable blockchain. Thus, Ethereum allows to create new applications on its blockchain and these applications are not limited to cryptocurrencies. A virtual node is able to run and execute small programs, called 'smart contracts'. Smart contracts run directly on the blockchain, as a result, they run exactly as programmed. Smart contracts denote how things need to be done in the Ethereum ecosystem, they basically are a set of instructions to perform a task. Hence, each dAPP consists of smart contracts.

Each dAPP has its own token, which the user can buy with Ether. Hence, this token is used to pay in the specified dAPP. There are hence currently thousands of tokens traded on the Ethereum blockchain. Note that the concept of a token is not new. Actually in real life we do the same thing. Suppose you go to an all inclusive party, where you buy the ticket with Euros and then the ticket gives you access to all the food and drinks at the party. The all-inclusive party is then the dAPP, Ether corresponds to Euros and the dAPP specific token is then the ticket.

2.2.2.1 Augur

A well-known applications which run on the Ethereum blockchain is Augur (Munster 2018; Augur 2018; Hussey 2018). Augur is basically an open-source decentralized prediction market and actually a good illustration of how Ethereum smart contracts can work. The system allows bettors from all over the world to place a bet on literally anything. It is founded by a computer scientist Joey Krug and software developer Jack Peterson. Augur consists of smart contracts which run on the blockchain. After it was first released, lots of people placed bets on it. However today, the activity has been limited, partially because the application is to some extent controversial. Bettors for example could bet on some kind of 'assassination market'. You win if for example a celebrity dies (or gets killed). People feared that bettors might actually kill celebrities if the jackpot is high enough. Augur is a betting website without a betting license and hence there is legal uncertainty around it (Team BEGN 2018).

Augur is based on the concept "**wisdom of the many**", it actually makes its predictions based on the opinion of the crowd. For example, farmers could hedge the loss of a bad farming season by exactly betting on that. Vitalik Buterin, the founder of Ethereum, is also a key advisor in Augur.

If someone wishes to create an Augur market, he or she has to follow 4 stages:

- **Creation**: to create a market, the creator has to spend a small amount of Ether and make a description of the market. The creators also have to choose how much it costs to bet on the market.
- **Trading**: this stage refers to the betting on the market created in stage one. Bettors actually buy shares in the outcome. A share becomes more valuable if its probability to become true becomes higher.
- **Reporting**: reporters are members of the network who verify a certain outcome and own the REP tokens (token involved in funding Augur). These reporters place a bet on how much they believe their reporting is true near maturity. If they

are right, they get a part of the jackpot. If they are wrong or do not bet, they lose REP.

- **Settlement**: once the reporting is complete and verified by the community. The bettors of the correct outcome are paid, as is the creator.

Augur predicted the outcome of the US midterm elections in 2018 correctly. It gave 82% probability that the Republicans would keep the senate and 65% chance that the Democrats will conquer the house (Munster 2018). The basic idea in this particular case is the many are stronger than political analysts and polls. Especially, if there is money involved for the people who are betting. However, the track record of Augur has been poor, handling with care is still needed.

2.2.3 Litecoin (LTC)

Litecoin has came to life in order to solve Bitcoin's major flaws. It is founded by Charlie Lee, an ex-employee of Google, on 7 October 2011 (BlockGeeks 2018).

Litecoin uses a Proof-of-Work algorithm called Scrypt (Bisola 2018), which is a password based algorithmic function to generate cryptographic keys. The transactions of Litecoin are cheaper and near-instant (Litecoin 2018). It generates blocks faster and therefore the network supports more transactions. The block generation takes approximately 2.5 min. Accordingly, the users receive confirmation faster. The reward for mining a block is halved every 840,000 Litecoins mined, which happens roughly every 4 years. In September 2019 the reward is 12.5 LTC. The faster block generation also provides more miners to claim the block reward. Theoretically, the block reward in Litecoin is spread more evenly over its users. Moreover, it has an improved storage based efficiency; Litecoin is able to handle an higher transaction volume than Bitcoin. Therefore, the average transaction fee is only 0.05 USD (Bit Info Charts 2019) and a transaction is fully confirmed in 30 min.

Litecoin is based on the same concepts as Bitcoin, it is a peer-to-peer, open source, global payment network (Moskov 2017). All the transactions are recorded on a distributed public ledger. The total number of Litecoin is capped to 84 million which is 4 times more than Bitcoin. Consequently, Litecoin is insensitive to inflation and deflation. One often says that Bitcoin is gold, used for storing value for long-term purposes while Litecoin is silver, used in transactions for cheaper and everyday purposes. One of the largest disadvantages of Litecoin is that it came after Bitcoin. When Litecoin was launched, Bitcoin already attracted most market participants. The market capitalization of Litecoin (CoinMarketCap 2019) is 4.486 billion USD in September 2019, which is much lower than Bitcoin's.

Scrypt is an alternative to SHA-256 in Bitcoin. However, in contrast to SHA-256, Scrypt is memory intensive since numbers are generated rapidly and they are stored in RAM, which needs to be accessed continuously. However, Scrypt has a lower hash rate than the algorithm used in Bitcoin. The main reason to use Scrypt was to avoid ASIC users to be superior. ASIC, Application Specific Integrated Circuit chips, users basically have specialized equipment to solve the cryptographic puzzle behind the mining process. ASIC hardware is designed with the particular hash algorithm in mind. Scrypt is a memory hard problem, hence, processing memory hard problems in parallel as ASIC users do, is extremely memory costly. Litecoin wants to give 'normal users' a chance to compete with the specialized users. Unfortunately, companies like Zeus and Flower managed to create ASIC Scrypts.

2.2.4   Monero (XMR)

Monero is a cryptocurrency which mainly focuses on non-traceability and privacy. This cryptocurrency resembles ordinary cash the most of all the cryptocurrencies in terms of privacy. It was created in April 2014 by Nicolas van Saberhagen.

The main feature of Monero is the blockchain obfuscation, the coins main concern is privacy. It starts by using the Cryptonote protocol, a Proof-of-Work consensus algorithm behind Monero. A miner receives 0.6 XMR (Gola 2018) per block mined. Monero also includes transactions fees, these are around 2 dollar on 2th of February 2019 (Monero 2019b). These transactions fees depend on the amount traded like in Bitcoin. The transaction confirmation time also depends on the number of transactions, however, on average a transaction is completely settled in 30 min on Kraken (a major exchange) (O'Keeffe 2018), this is after 10 blocks are added to the chain. Moreover, it takes on average 2 min before a new block is created (Monero.how 2017; BitInfoCharts 2019).

The market cap of total Monero is set at 300 million (BlockGeeks 2019). However, even if the supply of Monero coins runs out, there will be a continuous supply of 0.3 XMR/min to incentivize miners (BlockGeeks 2019). Monero has a total market capitalization of 1.264 billion US dollars on the 10th of September 2019 and is traded on all the major exchanges (Coinmarketcap 2019). The price of Monero is entirely determined by its popularity unlike Tether and Ripple for example.

The main features of Monero are ring signatures, stealth addresses and ring confidential transactions to ensure anonymity. Unlike Bitcoin, Monero is able to hide the identity of the sender, the receiver and the amount of coins involved in the transaction by grouping public keys together. This last feature is actually called ring signatures. In fact, Monero uses more than two keys to encrypt the data. Moreover, Monero is a reaction to transaction tracking, which is possible in Bitcoin. Monero is open-source and entirely permission-less, which means that anyone can access the platform for free.

In Bitcoin, addresses who have participated in undesired activities can be blacklisted. While Monero addresses are indistinguishable and thus cannot be blacklisted due to unwanted activities. This last features is made possible through the fungible property of Monero. This means that two coins of Monero are identical to its owner. For example, cars are not fungible (Monero 2019a). This means that you are not indifferent between owning two different cars. Suppose that you lend your car to your neighbour, then you want your car back and not some other car. A kilogram of salt, on the other hand, is fungible because you are indifferent in getting back your kilogram of salt or a new pack. This fungible quality becomes especially useful when you consider the request for clean coins. Which means that a specific coin cannot have been used for any illegal purposes. Moreover, in some cases these clean coins can be of higher value than their faulty colleagues (BlockGeeks 2019). In Monero, however, this problem does not occur, since it is not possible to track the history of one particular coin and thus every coin is evenly "clean".

One other major benefit of Monero is the dynamic scalability, this indicates how fast the network can grow in terms of demand. Unlike Bitcoin, Monero does not place a limit on the block size (King 2018). Suppose there is a limit like in Bitcoin, this means that transactions have to "battle" to be included in the block. Since miners will probably take on transactions which include higher transaction fees, the transactions fees will rise. Moreover, the duration to be included in a block can also increase significantly, because a new block is only mined every couple of min/s and its capacity to include new transactions is limited. In other words, due to the dynamic scalability of Monero, all these problems are solved. In order to prevent spam transactions, Monero has a block reward penalty system. It works as follows, take the median of the previous 100 blocksizes. If the blocksize of the new block is larger, the reward in mining decreases. Hence, including spam transactions in the block is gravely discouraged.

For many cryptocurrencies special mining groups are set up, these make it almost impossible for regular miners to compete. For example in Bitcoin, Application Specific Integrated Circuit chips (ASIC) are very popular, though expensive, mining operations. Monero uses CryptoNight to perform the hashes, this algorithm makes it a lot more expensive to set op ASICs and therefore such operations are unprofitable (King 2018; BlockGeeks 2019). In some sense, this allows Monero to be more decentralized than Bitcoin, because now CPUs (Central Processing Unit) and GPUs (Graphic Processing Unit) can also participate in mining blocks. However, in reality roughly 43% of the hashrates are controlled by only 3 mining pools in September 2019 (mineXMR 2019; BlockGeeks 2019; King 2018).

The main premise of Monero is its privacy and anonymity. In order to ensure the privacy of the recipient, every wallet has its own stealth address, this ensures unlinkability. A stealth address enhances privacy by creating a one-time address for every transaction of the sender to ensure that different payments are unlikable. On top of that, there is a built-in action called ring confidential transactions. This is an alternative to mixing in Bitcoin (see Sect. 3.4), which is the signing of transactions by a group of currency owners with the same amount of coins. In Monero, the exact amount is split up and hidden in the blockchain. In other words, it allows transactions to be anonymous. The privacy of the sender is ensured through ring signatures. Ring signatures actually group transactions, these are signed with identical signatures such that the actual sender cannot be retrieved. Monero uses key images to prevent double spending. Because in Monero everything is cloaked and hidden, every transaction comes with its own unique key image. Although privacy in this world is found very valuable it also has major downsides. Monero is the dark web its favorite currency as described in several news articles, due to its privacy features.

2.3 Byzantine Fault Tolerance Coins

2.3.1 *Ripple*

Ripple was established in 2012 by Ripple Company. Ripple is a so called payment infrastructure token. More specific it is a real-time gross settlement system, currency exchange and remittance network.

In contrast to Bitcoin, Ripple does not rely on the Proof-of-Work consensus algorithm which makes it less energy-draining and computational intensive. It uses an iterative consensus ledger together with independent validating servers which constantly compare transaction records (Shawn 2019). Ripple is designed to trade money nearly instantaneous (Ripple 2018). One of the cornerstones of Ripple is fast transactions rates, it has on average a block production which is 10 times faster than Bitcoin and 3 times faster than Ethereum. Payments in Ripple settle in only 4 s (Ripple 2018) with a transaction cost of approximately 0.004 USD.

One major difference with Bitcoin is that Ripple has been pre-mined (Cointelegraph 2019). Hundred billion XRP have been mined, where a part is available in the market and the remaining XRP are in Ripple Labs. Moreover, Ripple has a market capitalization of 11.265 billion Dollars (Market capitalization ripple 2018) on the 10th of September 2019. Ripple cuts out all the intermediaries in cross-border payments, it merely acts bank-to-bank.

RippleNet (Ripple 2018) is the solution provided by Ripple to solve the typically more slow settlements, with often high fees and unreliable standards for payment systems today. Ripple tries to connect banks and payment providers all around the world via RippleNet. It runs on a blockchain network and therefore offers near instantaneous settlements with a standardized access to the platform and lower capital requirements for cross-border payments. RippleNet uses XRP as an intermediary between several currencies to ensure liquidity on-demand. XRP is the in-house cryptocurrency of Ripple, it is especially designed for payments. The idea is that using XRP as an intermediary results in faster transactions, lower costs and more scalability than other digital assets. Prominent users of RippleNet are Santander, RCB, Crédit Agricole, . . . (Ripple 2018).

Ripple offers three specific solutions xCurrent, xVisa and xRapid. xCurrent aids banks in providing cross-border payments, while xRapid allows payment providers and other financial institutions to minimize liquidity costs. Lastly, there is xVisa which allows corporations, financial institutions and payment providers to send payments across numerous networks with a standard interface.

2.3.2 Stellar

Stellar offers a network which connects people, payment systems and banks on a global scale. Offering fast, reliable and cheap cross-border payments is the aim. It was released in 2014 by a nonprofit organization called Stellar Development Foundation. The main developers were Jed McCaleb and Joyce Kim. Note that Stellar is created by the same person as in Ripple, namely Jed McCaleb. However, Stellar focuses on facilitating payments between individuals and the developing world, although it can also be used by banks (King 2018).

Transactions in the Stellar network are verified in only 2–5 s (Stellar Development Foundation 2017) by means of the BFT consensus algorithm. Each Stellar server communicates and syncs with each other to ensure that every transaction is valid and becomes part of the global ledger. A transaction costs 0.00001 XLM, where XLM stand for Lumens, the in-house currency involved in Stellar.

Initially the amount of Stellar was capped to 100 billion (Munro 2018), however, now the supply is unlimited. This ensures that Stellar is an inflationary currency with a fixed inflation rate of 1% a year (King 2018). Moreover, all transaction fees are recycled. Note that Stellar cannot be mined. Stellar has a market capitalization of 1.159 billion USD (Coinmarketcap 2019) in September 2019.

The Stellar network (Stellar Development Foundation 2017) consists of decentralized (independent) servers spread over the world to power a distributed ledger. The decentralization ensures that the network will run successfully even if some servers fail. Moreover, a copy of the global ledger, which records balances and transactions, exists on every Stellar server.

Stellar has a distributed exchange. This means that if you send Euro credits to someone in the United States of America with a Dollar credit balance, the network converts the amount of Euro credits to Dollar credits at the best available rate possible. The Stellar ledger can help people buy other currencies. It contains an order book which presents all the different buy and sell prices of all the different currencies. Hence, the converting of currencies actually works by using this order book. There are three ways to convert currencies (Munro 2018), namely:

- **Peer-to-Peer**: look for people who want to make the opposite trade of currencies directly without intermediary of XLM.
- **Via XLM**: use the native currency of Stellar as an intermediary.
- **Chain of Conversions**: the network converts into different currencies in a row to reach the best exchange rate possible.

Note that the Stellar network chooses automatically the best option to perform the transaction.

Anchors hold the deposits and issue credits in the network, anchors are entities not humans. They merely acts as a bridge between different currencies and the Stellar network. Anchors have two different jobs:

- Receiving the deposit and issue the corresponding credit to the account address on the ledger;
- Giving the amount withdrawn by subtracting the corresponding amount of credit on the balance.

Anchors can be compared to banks and payment networks.

Not only does Stellar facilitates payments around the world, it also provides development friendly software and tools (Stellar Development Foundation 2017). Users of Stellar can build financial services on the Stellar network. It is possible to build mobile wallets, online banking apps and basically every application concerning payments. Stellar has forged a partnership with IBM (King 2018).

2.3.2.1 Ripple Versus Stellar

Stellar is actually a close competitor of Ripple, since both offer fast, reliable and affordable global payments. Both also want to provide cheap transfers. Ripple focuses on improving the efficiency between the different financial institutions and companies, while Stellar its centre of attention is providing low-cost financial services. In other words, they both address different aspects of the market. On top of that, Stellar also offers a platform for ICO launches and a decentralized exchange. Ripple is not decentralized since it uses banks in order to perform payments. Another big difference is that Ripple is a privately-owned company which sells its product to other companies, while Stellar is a non-profit organization.

In Ripple, the financial institutions and multinationals use the native token (XRP) to make international payments, which ensures liquidity on demand. Stellar focuses more on individuals who directly trade with one another. Stellar uses Lumens (XLM) as a medium to handle the fiat currency aspects.

Ripple uses the proof-of-correctness consensus mechanism based on majority validation, while Stellar uses the Stellar consensus algorithm. The consensus algorithm in Stellar is based on the Federated Byzantine Agreement, this ensures that consensus can be reached without relying on a closed system. The supply availability of Stellar and Ripple is also different, Stellar offers an unlimited supply in contrast to Ripple. Moreover, a small amount of XRP is destroyed in each transaction in order to protect the network against spam and denial-of-service attacks. Therefore, the total supply of Ripple will decrease over time.

At this point in time, Ripple has forged more important partnerships and its main goal for the future is to be adopted by financial institutions as the new Visa. Ripple is also currently larger than Stellar, however if Stellar is able to secure enough users around the world, its transactions could become be cheaper.

2.4 Proof of Stake Coins

2.4.1 EOS

EOS is a decentralized operating system based on blockchain technology and is first mentioned in a whitepaper published in 2017. On the first of June 2018, EOS was officially launched on an open-source platform. Block.one, a private owned company, is the developer behind EOS. This digital coin is similar to Ethereum, unofficially EOS stand for Ethereum Operating System. Therefore EOS is also a general platform token. The cryptocurrency was made possible through an ICO of nearly 4 billion USD (Wilmoth 2018), which is one of the biggest ICO's until now.

The Delegated Proof-of-Stake algorithm, together with the fact that EOS runs on various computer cores, allows it to eliminate transaction fees (Brian 2018). In order to cover the cost of running the network, inflation is used. 1% of the yearly inflation is allocated to the block producers to secure the network. In other words, stakeholders are paying fees through increased supplies and not the end users as is the case in Ethereum. EOS offers a much greater capacity, with up to a 4000 transactions per second (Block.one 2019) (Ethereum can only do 15 transactions per second Macdonald 2018). A new block is 99.9% valid after only 0.25 s due to a signing system between the block producers.

Upon the release, block.one sold one million tokens on an Ethereum platform to ensure a wide distribution of the native EOS tokens, more EOS tokens cannot be mined. These tokens are used to buy both storage and bandwidth on the EOS blockchain. However, a token allows the owner to participate in on-chain governance and to cast votes. Although, EOS only exists from 2018, it has a market capitalization of 3.387 billion USD on 10 September 2019 (Coinmarketcap 2019).

Now we will provide more evidence on the block production algorithm in EOS. Blocks are produced in rounds of 126 by using 21 producers, who need to be elected through the Delegated Proof-of-Stake algorithm (Block.one 2018). If a fork occurs, the consensus algorithm will switch to the longest chain. Moreover, block producers do not compete, they cooperate, therefore it is highly unlikely that a fork even occurs. A block producer, who does not act fair and honest, will probably be voted out.

The on-chain governance system will lead to more politics because each token holder has a vote. In EOS the majority of block producers can change the constitution behind EOS. If 15/21 vote for the change, then the constitution can be changed. However, block producers need to be elected by the token holders, this voting system is known as the Delegated Proof-of-Stake algorithm. EOS prefers this algorithm over the Proof-of-Work algorithm, used in Ethereum, to ensure a better performance. However, this algorithm also makes EOS more centralized.

EOS supports dAPPs on a commercial-scale, meaning that apps can be developed similar to web-based applications. If a particular dAPP needs 10% of the total bandwidth of EOS, the creator of the dAPP needs 10% of the total EOS available. Hence, dAPPs actually secure the network through the inflation (Macdonald 2018).

In time as all EOS tokens have been sold, owners can rend their tokens to creators of dAPPs. The price of an EOS token will grow linearly with the network adoption. Another big advantage for EOS its users is that there is a recovery algorithm which allows the user to retrieve lost private keys again. Also, the user names in EOS are human-readable. For example, a consumer can now send coins to @ElineVanderAuwera rather than the address used in Ethereum which is a string of seemingly random characters.

2.4.2 Tether (USDT)

Tether is different from most other cryptocurrencies, it is actually pegged to a real-world currency and therefore called a "stable coin". Tether was originally announced under the name "Realcoin" in July 2014 by the co-founders Brock Pierce, Reeve Collins, and Craig Sellars as a Santa Monica based startup. The Tether whitepaper was already published in 2012.

Five different blockchains support Tether, namely the platform of EOS, Tron, Ethereum, Omni (Bitcoin blockchain) and Algorand (Zmudzinski 2019). Therefore, Tether has two different consensus algorithms, namely PoW and PoS. There is no transaction fee between two Tether transactions (Tether 2018), however there is a fee if there is a fiat withdrawal or deposit as shown in Table 2.1. Moreover, there is a 150 USD verification fee to open up a tether account (Tether 2018).

Despite having a rather low market capitalization of 4.086 billion Dollars in September 2019, it is one of the most traded coins. Mainly because many exchanges offer Tether as a trading pair (CryptocurrencyFacts 2019). In other words, user can buy other cryptocurrencies with a coin that basically represent the USD.

Before 2019, Tether is said to have an one-on-one relation to the Dollar, however, it also supports the Euro and soon the Japanese Yen (Tether 2018). From 2019 onward, Tether is said to be 100% backed by its reserves. Because Tether is 'tethered' to real-world currencies, it is supposed to have both advantages of cryptocurrencies and real-world currencies. For instance, it does not suffer the same volatility issues as other cryptocurrencies, it is in fact designed to be stable. It is able to move across a blockchain, which is faster, more anonymous and cheaper than regular currencies. Tether is not money, it still is a digital token which runs on a blockchain and therefore it is not really equal to the Dollar. Tether allows

Table 2.1 Fees included in Tether transactions

30 day transaction value sum range	Fee per fiat withdrawal	Fee per fiat deposit	Fee per Tether deposit or withdrawal
$100,000–$999,999	max ($1000 and 0.4%)	0.1%	Free
$1,000,000–$10,000,000	1%	0.1%	Free
≥$10,000,000	3%	0.1%	Free

the user to store, receive and send tokens globally, instantaneously and securely with fewer costs involved than the current alternatives. Tether coins are issued by Tether Limited, a company which under the governance of the British Virgin Islands. Many buyers of Tether intend to use the coins to buy other cryptocurrencies. For this reason, the crypto economy is "tethered" to a substitute of the Dollar, which value cannot be guaranteed.

Tether promises its customers near zero conversion fees, no commission and top market exchange rates (Tether 2018). Moreover, Tether promises to operate in a transparent and secure way, while trying to adhere all the governmental regulations regarding cryptocurrencies. In order to do so, Tether publishes its reserve holdings daily and audits them frequently. It ensures that its assets match the total amount of Tether in the market at any time.

Many news sites have raised their concern regarding Tether. Bloomberg states that crypto traders are worried about Tether (Tan et al. 2018) and CNBC quotes that Tether could have devastating effects on the market (Kharpal 2018). According to CNBC, Tether is used to buy up cheap Bitcoins. During December 2017 and January 2018 the price of Bitcoin has reached tremendously high levels after which they dropped. At the same time, Tether Limited has been releasing over 850 million coins onto the market (Kharpal 2018). An analysis has also proven that a creation of Tether often coincides with a drop in the price of Bitcoin is falling (CryptocurrencyFacts 2019). Tether and Bitfinex, one of the world's biggest cryptocurrency exchanges, have the same CEO. Therefore, trading in Tether can also prop up the price of Bitcoin. CNBC and Bloomberg also question the fact that Tether is fully backed by the Dollar. On the 15th of October 2018 the value of Tether, which was always strict around 1 Dollar, fell to 90 cents.

2.5 Toy Coins

Many coins which were founded after Bitcoin and many in the meantime seized to exist. Some coins even were a scam from the beginning, just to extract money from innocent investors. One example is Kitty Coin (Dead Coins 2018). The DEVs abandoned development, the cord was deleted and the website disappeared from one day to the other. An example of a dead coin is Secretcoin, the developers left in May 2018.

Needless to mention, there are a lot of coins in the market. There exist many coins which are developed just for fun, such as Jesuscoin, Satancoin, Shitcoin, Dogecoin, ... Clearly, any self respecting investor would opt not to invest in any of these coins simply because investors do not believe the coin represents any seriousness. However, Dogecoin managed to develop an internet community willing to invest in it. It even reached a market capitalization of 450 million USD (Coinranking 2018) on October in 2018.

Dogecoin (Zoe 2018; Dogecoin 2018) is a "joke currency" developed at December 6, 2013 by Billy Markus and Jackson Palmer. Originally there was a cap on the amount of Dogecoin to be mined of 100 billion coins. Later one of the founders removed this limit. One particular reason is that the community behind Dogecoin wanted a coin that could be used on a daily basis. It is one of the most circulated coins, mainly because it is used in an internet tipping system and it has low transaction fees. The Dogecoin has become a strong altcoin with lots of potential due to the strong community behind it. The community has done special things with Dogecoin, not only did they said the Jamaican Bobsled team to compete in the winter Olympics but they also invested in a well in Kenya. Currently, it is the top 25 cryptocurrencies.

It is built using the Bitcoin code base but it is also based on Litecoin and on top of that it offers an high liquidity. Dogecoin can be mined and this process is much easier and faster than Bitcoin. A new block is created on average in only 1 min in comparison to the 10 min of Bitcoin. Moreover, Dogecoin rewards a fixed amount of coins for successfully mining a block, like Ethereum.

2.6   Summary of the Key Characteristics (Table 2.2)

Table 2.2 Most important exchanges according to trading volume on 10 September 2019

	Consensus algorithm	Block reward	Block creation time	Supply	Market cap. (billion USD)	Transaction fee (USD)	Transaction confirmation time
Bitcoin	PoW	12.5 BTC	10 min	21 million	184.566	0.50	60 min
Ethereum	PoW	5 ETH	14–15s	Unlimited	19.645	0.12	6 min
Ripple	BFT like			100 billion	11.265	0.004	4 s
Litecoin	PoW	12.5 LTC	2.5 min	84 million	4.486	0.03	30 min
EOS	dPoS	1% of yearly inflation	500 ms	1 million	3.387	0	0.25 s
Stellar	BFT like			Unlimited	1.159	0.064	2–5 s
Monero	PoW	0.6 XMR	2 min	300 million + 0.3 XMR/min if supply runs out	1.264	2	30 min
Tether	PoS/PoW			Unlimited	4.086	Table 2.1	Depending on the blockchain

References

Augur (2018) A decentralized oracle and prediction market protocol. https://www.augur.net/
Barber S, Boyen X, Shi E, Uzun E (2012) Bitter to better—how to make bitcoin a better currency. In: Keromytis AD (ed) Financial cryptography and data security, vol 7397. Springer, Berlin, pp 399–414
Bisola A (2018) Litecoin scrypt algorithm explained. https://www.mycryptopedia.com/litecoin-scrypt-algorithm-explained/

Bit Info Charts (2019) https://bitinfocharts.com/

Bitcoin.com (2019) Differences between bitcoin and bitcoin cash. https://www.bitcoin.com/info/differences-between-bitcoin-cash-bch-and-bitcoin-btc

Bitcoin.com (2017) What is bitcoin? https://www.bitcoin.com/info/what-is-bitcoin

BitInfoCharts (2019) Monero block time. https://bitinfocharts.com/comparison/monero-confirmationtime.html

BlockGeeks (2019) Guide to monero. https://blockgeeks.com/guides/monero/

BlockGeeks (2019) What is cryptocurrency: everything you must need to know! https://blockgeeks.com/guides/what-is-cryptocurrency/

BlockGeeks (2018) What is Litecoin? The most comprehensive guide ever! https://blockgeeks.com/guides/litecoin/

Block.one (2018) Eos.io technical whitepaper. Technical report, Block.one

Block.one (2019) Block.one releases EOSIO 1.6.0, sees potential 35% increase in transaction speeds

Brian P (2018) What is EOS? Introduction to the EOS token. https://coinswitch.co/info/eos/what-is-eos

Coinmarketcap (2018) Market capitalization ripple (2018) https://coinmarketcap.com/currencies/ripple/

Coinmarketcap (2019) Market capitalisation of cryptocurrencies. https://coinmarketcap.com/

Coinmarketcap (2019) https://coinranking.com/

Coinranking (2018) https://coinranking.com/coin/dogecoin-doge

Cointelegraph (2019) Ripple vs. Bitcoin: key differences. https://cointelegraph.com/ripple-101/ripple-vs-bitcoin-key-differencesmining-rewards

Cryptocurrency Army (2019) Properties of cryptocurrencies. https://www.cryptocurrencyarmy.com/properties-of-cryptocurrencies/.

CryptocurrencyFacts (2019) What is tether? https://cryptocurrencyfacts.com/what-is-tether/

Das S (2018) Banks Begone: UN's world food programme builds on Ethereum blockchain money transfers. https://www.ccn.com/banks-begone-uns-world-food-programme-builds-ethereum-blockchain-money-transfers

Dead Coins (2018) A complete list of ICO exit scams and extinct coins. https://deadcoins.com

Dogecoin (2018) https://dogecoin.com/

Elliott DJ, de Lima L (2018) Crypto-assets: their future and regulation

Elliott, DJ, de Lima L, Singel R (2018) Cryptocurrencies and public policy: key questions and answers. Technical report, Oliver Wyman

Ethereum Foundation (2018) Ethereum project. https://www.ethereum.org/

Glaser F, Zimmermann K, Haferkorn M, Weber M, Siering M (2014) Bitcoin—asset or currency? Revealing users' hidden intentions. In: Proceedings of the 22nd European Conference on Information Systems, ECIS 2014

Gola Y (2018) Over 90 of moneros block reward has been mined. https://www.ccn.com/over-90-of-moneros-block-reward-has-been-mined/

Grinberg R (2011) Bitcoin: an innovative alternative digital currency. Hastings Sci Technol Law J 4:160

Half BB (2019) Bitcoin block Halving of reward countdown. https://www.bitcoinblockhalf.com/

Hussey M (2018) Intro to augur. https://litepaper.com/resources/intro-to-augur?utm-source=decrypt-media&utm-media=embedded-widget&utm-name=decrypt-media-embedded-widget

Kahan M, Rock E (2008) The hanging chads of corporate voting. Georgetown Law J 96(4):1227–1281

Kharpal A (2018) All you need to know about tether, the cryptocurrency that could cave devastating effects on the market. https://www.cnbc.com/2018/02/02/tether-what-you-need-to-know-about-the-cryptocurrency-worrying-markets.html

King R (2018) Monero. https://www.bitdegree.org/tutorials/monero/

King R (2018) Stellar vs ripple: what is the better choice? https://www.bitdegree.org/tutorials/stellar-vs-ripple/

Lewis A (2015) A gentle introduction to Ethereum. https://bitsonblocks.net/2016/10/02/gentle-introduction-ethereum/

Litecoin (2018) What is litecoin? https://litecoin.org/

Macdonald A (2018) EOS vs Ethereum: predicting the winner of the smart contract war. https://cryptobriefing.com/eos-ethereum-smart-contract-war-winner/

Mersch Y (2018) Virtual or virtueless? The evolution of money in the digital age. Lecture by Yves Mersch, Member of the Executive Board of the ECB, Official Monetary and Financial Institutions Forum

mineXMR (2019) https://minexmr.com/pools.html

Monero (2019a) Secure, private, untraceable. https://www.getmonero.org/

Monero (2019b) Coin news, guides and reviews. https://monero.org

Monero.how (2017) How long do Monero transactions take. https://www.monero.how/how-long-do-monero-transactions-take

Moskov A (2017) Litecoin vs. Bitcoin: comparing two of the most popular cryptocurrencies. https://coincentral.com/litecoin-vs-bitcoin/

Munro A (2018) XRP vs. Stellar: a side-by-side comparison. https://www.finder.com/ripple-vs-stellar-lumens

Munster B (2018) Augurs prediction: GOP loses the house but keeps senate. https://decryptmedia.com/2018/11/05/augurs-prediction-gop-loses-the-house-but-keeps-senate/

Munster B (2018) The long shot. https://decryptmedia.com/2018/10/15/augur-ico-prediction-market/

O'Keeffe D (2018) Understanding cryptocurrency transaction speed. https://medium.com/coinmonks/understanding-cryptocurrency-transaction-speeds-f9731fd93cb3

Ripple (2018) One frictionless experience to send money globally. https://ripple.com/

Rosic A (2017) https://blockgeeks.com/guides/what-is-ethereum-classic/

Rosic A (2018) What is an Ethereum token: the ultimate beginners guide. https://blockgeeks.com/guides/ethereum-token/

Satoshi N (2008) Bitcoin: a peer-to-peer electronic cash system. Technical report, bitcoin.org

Selgin G (2015) Synthetic commodity money. J. Financ Stability 17:92–99

Shawn G (2019) What is ripple? https://bitcoinmagazine.com/guides/what-ripple/

Stellar Development Foundation (2017) https://www.stellar.org/

Tan A, Robertson B, Leising M (2018) Why crypto traders are so worried about tether. https://www.bloomberg.com/news/articles/2018-10-14/why-crypto-traders-are-so-worried-about-tether-quicktake-q-a

Team BEGN (2018) 152 million dollar augur (rep) cofounder lawsuit dismissed, settled out of court. https://bitcoinexchangeguide.com/152-million-augur-rep-cofounder-lawsuit-dismissed-settled-out-of-court/

Tether (2018) Fees in tether. https://tether.to/fees/

Tether (2018) https://tether.to

Vasu-Devan H (2017) What is bitcoin and how does cryptocurrency work? (A primer). https://www.ngpf.org/blog/current-events/cryptocurrency-bitcoin-primer/

Vitaris B (2017) Swiss crypto valley to create digital identities for its citizens on the Ethereum blockchain. https://bitcoinmagazine.com/articles/swiss-crypto-valley-create-digital-identities-its-citizens-ethereum-blockchain/

Wilmoth J (2018) EOS ICO approaches 4 billion after year-long crowdsale. https://www.ccn.com/eos-ico-approaches-4-billion-after-year-long-crowdsale/

Yermack D (2015) Corporate governance and blockchains. Working Paper 21802, National Bureau of Economic Research

Zmudzinski A (2019) https://cointelegraph.com/news/tether-usdt-is-launching-on-a-pure-pos-blockchain-algorand.

Zoe B (2018) What is Dogecoin. https://www.investinblockchain.com/what-is-dogecoin/

Part II
Risk in Dealing with Cryptocurrencies

Chapter 3
Qualitative Risks

Abstract Blockchain shifts the trust in governments and centralized systems to trust in computers, cryptography, software and protocols. However, it is not because mathematics and cryptography are behind cryptocurrencies and blockchain, it necessarily means that all of the operational risks are eliminated and the system is more secure. For example, storing digital currency on an online wallet makes it prone to hacks and there is no way to retrieve money which has been stolen. Moreover, if the owner of the wallet for some reason loses the password, then it is also impossible to retrieve the lost funds. This next chapter is dedicated to explaining some new risks in cryptocurrencies and the important underlying characteristics one should bear in mind. We start by exploring the cryptocurrency market and regulatory issues. Afterwards, we will elaborate on design risks, like the 51% attack, selfish mining and ICOs. To finish this chapter, we will go through various operational risks.

Keywords Cryptocurrency market · First mover advantage · Market capitalization · Trend · Regulation · Know your customer · Anti-money laundering · 51% attack · Selfish mining · Initial coin offering · Wallet · Exchange

3.1 Cryptocurrency Market

Bitcoin was the first cryptocurrency and many cryptocurrencies still follow trends set out by Bitcoin. This particular cryptocurrency has a dominant position in the market, due to the first mover advantage. Figure 3.1 shows the market share of Bitcoin in blue (Coinmarketcap 2019). Until mid 2017, Bitcoin had more than 50% of the total market capitalization, now other cryptocurrencies are securing their share of the market, with Ripple and Ethereum as leaders.

The market capitalization of cryptocurrencies has grown over the past years as people became more and more familiar with the concept. Figure 3.2a represent the market capitalization of the whole cryptocurrency market. It seems like the market capitalization follows an exponential trend. Therefore, we take the logarithm of the

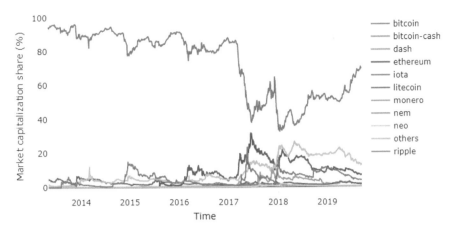

Fig. 3.1 Market share in percentage of the most prominent cryptocurrencies throughout time

market capitalization and check if this grows linearly. Figure 3.2b shows the market capitalization of the cryptocurrency market in the logscale. Overall, the market capitalization follows a linear trend in the logscale, which means that the market grows at an exponential rate $\sim \exp(kt)$ with $k = 0.002$ per day.

In Fig. 2.2 one can see how the prices of Bitcoin evolved over time. The price of Bitcoin remained fairly stable until it seemingly exploded in 2018. Afterwards, the price of Bitcoin has had bigger ups and downs than before. Figure 3.3 shows the trading volume of Bitcoin over time. From mid 2017 onward, the trading in Bitcoin increased rapidly. It seems like the trading volume of Bitcoin follows the trend of the price of Bitcoin. When the prices of Bitcoin began to fall in 2018 (see Fig. 2.2), the trading volume of Bitcoin also decreased a lot. Also, in 2019 the prices of Bitcoin were rising again and so is the trading volume.

3.2   Regulatory Issues

Governments all around the world try to get a grasp on the new currencies. However, putting regulations in place can be challenging due to a number of reasons. First and foremost, cryptocurrencies are spread in a global network and mined periodically to prevent any governmental control. It is hard for one nation to set up rules on its own. Several nations need to work together if they wish to set up an accurate and actually working legal framework without arbitrage opportunities. Moreover, cryptocurrencies are unlike regular currencies, managed by the network, hence not by any central bank or government.

A second reason is that a governance model is typically already in place in cryptocurrencies. Users of the network can decide how the network evolves and how to fix certain errors. However, in most networks, each user has an equal amount of

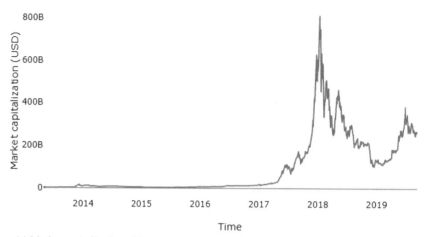

(a) Market capitalization of the cryptocurrency market as a whole

(b) Market capitalization in logscale with corresponding trend

Fig. 3.2 (**a**) Depicts the market capitalization of the whole cryptocurrency market in USD. (**b**) Shows the logarithm of (**a**) together with the trend of the market

voting power, hence, pre-existing governance models do not really apply. There are two types of governance models in cryptocurrencies; "on-chain" and "off-chain". On-chain mechanisms change the protocol based on rules and voting mechanisms already embedded in the protocol, while off-chain mechanisms can add new stake-holders and influencers in the governance procedure. These new types of governance offer new ways to deal with risk, but also introduce new risks. Users of the network can decide, in case of a hack, to adopt the chain without the hack. One of these new risks is for developers, who can be held accountable for a bug or an error in the code of which hackers made advantage. The underlying governance model proves that

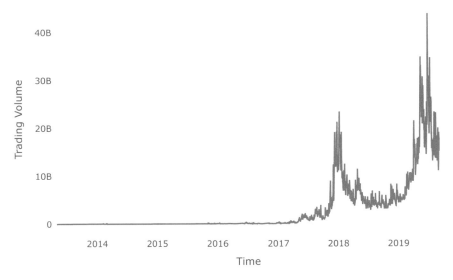

Fig. 3.3 Trading volume of Bitcoin (BTC) in USD (Coinmarketcap 2019)

cryptocurrencies are not fully decentralized, in the end there are always humans in charge. The rules, which seems to be set in stone, can always be altered by creating a hard fork. Even if the creators implement regulations, albeit demanded by the regulator, the community can choose to create a fork in which these regulations do not apply (Kroll et al. 2013).

A third reason is that it is hard to adapt pre-existing regulations to cryptocurrencies, because cryptocurrencies have features of three broad categories. They can be seen as a security, a commodity or a currency.

Some people are in favor of regulation others are not. Regulation on cryptocurrencies could attract different groups of investors, who now believe these assets are too "dangerous" to invest in or because of their unregulated character can just not be part of an investment portfolio. Many policy makers today are following a "wait and see" approach to observe the effects on the system and the evolution of the technologies in order to adapt regulation accordingly (European Commission 2018). Some regulators have already banned the use of credit cards to buy cryptocurrencies or even the exchange of cryptocurrencies altogether.

After the announcement of the Libra, regulators around the world have been speeding up their initiatives and have become actually more concerned. This means that cryptocurrencies are at risk to governmental decisions and this may very well be one of the greatest risks. Very targeted and coordinated regulation can cause steep declines in the value of cryptocurrencies.

China, on the other hand, proposed its own digital currency on 6 September 2019 (Holmes 2019). The coin will bear similarities to Libra and will help China keep its foreign exchange sovereignty. The deputy director of the people bank of China states that the digital currency will be as safe as central bank issued paper notes.

3.2.1 Regulation in Place

Brito et al. (2014) provide an overview of the regulation which was already in place prior to 2014 and the regulation they expect will come regarding cryptocurrencies. Before 2014, the main regulation on this subject concerned customer protection and money transfers. In the United States businesses which exchange cryptocurrencies, are considered as money transmitter businesses (Pieters and Vivanco 2017). As a result, these companies are required to comply with know-your-customer (KYC) regulation and must put anti-money-laundering (AML) schemes in place. On top of that, they are obliged to report suspicious behaviour. In Europe, the ECB (Mersch 2018) takes an active role in regulating or trying to regulate cryptocurrencies. It has voiced its opinion of the 5th anti-money laundering directive, which will extend the scope of obliged entities to adhere the KYC and AML directive to exchanges and wallets handling cryptocurrencies.

The KYC refers to the ability to retain and store sensitive information about a customer to reduce the risk for companies to be used in illegal activities. The AML regulation tries to avoid illegally obtained money to re-enter the financial system. These regulations limit the amount of illicit activities such as tax evasion, black market transactions, financing terrorism, avoiding sanctions, enabling ransomware payments and money laundering schemes.

Brito et al. (2014) also stress the implications of the category cryptocurrencies will fall into (commodity, currency, financial asset). Depending on the category, cryptocurrencies will be treated differently when implemented in a financial product (like a future, forward, ...). Financial regulators can opt to exclude certain financial transactions in cryptocurrencies from the full scope of the commodity Exchange Act regulation, similar to private security offerings and forward contracts. The authors also claim that too much regulation can outweigh the benefits of cryptocurrencies.

In 2017, Pieters and Vivanco (2017) state that cryptocurrency exchanges still do not have a global regulatory framework. They find that cryptocurrency exchanges that do not adhere the KYC regulation, exhibit different prices from exchanges which do.

3.3 Design Risk

Cryptocurrencies merely exist in the digital world, they are entirely programmed and coded. For this reason, bugs in the coding or technology specific problems are warp and weft. It is often the case that the programmers behind the cryptocurrency or exchange, even if these are fully decentralized, are held accountable for the errors or loss of money. One small programming error can lead to a loss of millions; people will always try to exploit errors. Some opt therefore to make the project open-source. This way other programmers can try to amend the code and make it safer for use. Moreover, coding errors can lead to operational risk, for example Binance suspended trading for almost 2 days to fix the bug in the code. The next two subsections explain two cryptocurrency specific problems related with the particular design of cryptocurrencies.

3.3.1 51% Attack

A 51% attack refers to one individual or a group, who seek to control over 50% of the total hashing or computing power in a certain cryptocurrency (Frankenfield 2019). Remember that many blockchains are governed by a majority, hence, if the majority of hashing power is in the hands of one single group of people, then they are able to control the course of action of the blockchain. In such an attack, the hackers are able to prevent certain transactions and reverse transactions that have already happened, which have not yet been confirmed. The former can bring the cryptocurrency in discredit, may lead to a decrease in value and disrupts the operation of the payment network (Rosenfeld 2014). The latter can enrich the attackers and is also known as double spending. In other words, first the attacker is able to convince the merchant that the transaction took place and then convinces the network that the transaction never took place or that another transaction happened (Rosenfeld 2014).

Kroll et al. (2013) argue that a cartel only has a limited payoff for executing a double spend attack and therefore cartels will not perform them. If cartels perform a lot of double spend attacks, the other players will loose faith in the value of the currency and the system and will not transact in it. As a result, the value of the currency would diminish and therefore the payoff of a double-spend attack becomes smaller. Thus, the cartels could destroy the currency they invested in by performing a double spend attack.

In case of a 51% attack, the group of fraudsters are able to built a chain at a faster rate than the honest majority (Goyal 2019). At some point, the faulty blockchain will become the longest chain in existence. As part of the attack, this chain will then be broadcasted to the network only when the merchant believes his transaction is confirmed. The network, on its turn, will perceive this chain as the legitimate chain, since it is the longest. Hence, it will be accepted by all the participants and everybody continues building on it. Therefore, all transactions that are not included in this faulty chain are considered wrong and are reversed. For example, if someone bought a car and received his car using the actual legitimate chain and then initiated a 51% attack, where he did not include the transaction of buying the car, then the person would own the car and the coins he used to pay for the car. In fact, double spending can be executed against multiple participants.

The 51% attack is mainly a problem in smaller cryptocurrencies which use the
Proof-of-Work consensus algorithm, because it is easier to take over the majority of
the computing power. In general, these kind of attacks are rare because one would
need to have state-of-the-art computers who can compete with the rest of the world
to create the longest chain (S 2018). Moreover, the electricity and storage cost for all
the mining hardware and covering the tracks can also not be neglected in performing
a 51% attack.

The probability of a successful double spending attack depends on the number
of blocks and the number of confirmations needed for a transaction to be final
(Rosenfeld 2014; Goffard 2018). The number of blocks created by the attacker (m)
is distributed according to a negative binomial distribution (Rosenfeld 2014; Goffard
2018). We will denote the number of blocks created by the honest network as h and
the probability that the attacker creates a block is q. Then

$$P(m) = \binom{m+n-1}{m}(1-q)^n q^m. \tag{3.1}$$

The probability that a double spend attack succeeds conditional on the fact that
the merchant waits for n blocks to be mined before executing the transaction is
Rosenfeld (2014):

$$\sum_{i=1}^{m} P(m)a_m \tag{3.2}$$

where a_m is the probability that the attacker(s) are able to build the chain at a faster
pace and are thus able to catch up to the legitimate chain when they are m blocks
behind. It is possible to write a_m as a recurrence relationship

$$a_m = (1-q)a_{m+1} + qa_{m-1} \tag{3.3}$$

Combining Eqs. (3.2) and (3.3) gives:

$$\begin{cases} 1 - \sum_{m=0}^{n} P(m) - \binom{m+n-1}{m}(1-q)^m q^n & \text{if } q < \frac{1}{2} \\ 1 & \text{if } q \geq \frac{1}{2}. \end{cases} \tag{3.4}$$

Moreover, Eq. (3.2) clearly shows that the probability of a successful 51%-attack
depends on the number of blocks and the number of confirmations needed.

As mentioned before the 51% attack mostly affects smaller cryptocurrencies,
however, on the 18th of May 2018 Bitcoin Gold, a hardfork of the original Bitcoin,
suffered a 51% attack (Osato 2018). Bitcoin gold is the 28th largest cryptocurrency
in terms of market capitalization (Coinranking 2019). Figure 3.4 represents the
presumable address of the hacker. If all of the transactions listed in this figure are

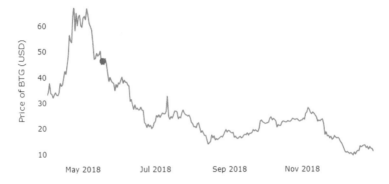

BTG Address

Summary			Transactions	
Address	GTNjvCGssb2rbLnDV1xxsHmunQdvXnY2F1		No. Transactions	76
BTC Format	1AXpW4wvtjRZWsUv75JrSXS1sLr57sUaUc		Total Received	388,201.92404001 BTG
Final Balance	12,239.00 BTG		Total Send	375,962.92404001 BTG

Fig. 3.4 Bitcoin Gold address of the suspected hacker

Fig. 3.5 The Price Evolution of Bitcoin Gold. The 51% attack happened on the 18th of May 2018 and is represented by the red dot

part of the hack, then almost 18 million USD worth of Bitcoin Gold is stolen. As long as the hacker holds over 51% of the computing power, he is able to continue the attack. However, in this particular case all transactions seized the next day. According to the Bitcoin Gold communications director, there is no risk for normal participants as long as they do not accept large amounts of payments. As a consequence of such an attack, the integrity of Bitcoin Gold was severely affected. As expected the price dropped after the attack, as can be seen in Fig. 3.5. There are in fact numerous other currencies who are a victim of a 51% attack; NEM, Verge and ZenCash to mention some (Copeland 2019).

Note that a 51% attack is not a bug, it is an intrinsic feature of decentralized blockchains, since a blockchain is controlled by the masses in order to be decentralized. One possible solution to prevent an attack is to penalize miners who broadcast long chains to the network. Another resolution is given by using delayed Proof-of-Work introduced by Komodo (2018). Basically, Komodo stores backups of its blockchain on the Bitcoin ledger. It achieves this by making a small transaction in the Bitcoin blockchain. Every 10 min a block hash from a block in the Komodo network is written into a block on the Bitcoin blockchain by spending a small amount of Bitcoin to record the transaction with the hashed info on the Bitcoin blockchain. This process is called notarization and it provides the same level of security as the Bitcoin network. In this particular security mechanism, the

51% attack needs to happen in the 10 min between the production of two blocks. Furthermore, delayed Proof-of-Work has been applied to other blockchains, like HUSH, Kreds, GAME Credits, ... This technique enables small blockchains to be more secure and safe for usage.

3.3.2   Selfish Mining

Many authors (Kroll et al. 2013; Schrijvers et al. 2017) believe that the blockchain protocol in Proof-of-Work cryptocurrencies is incentive-compatible. This means that all the players have the incentive to act honest and follow the protocol of the blockchain. However, Eyal and Sirer (2018) and Sapirshtein et al. (2017) have found a strategy which shows that it is preferred to perform this strategy since it offers a proportionally higher payoff.

Selfish mining was first introduced by Eyal and Sirer in 2018. In selfish mining, a pool of miners receives a return which is proportionally higher than the effort (mining power) invested. In case of a successful selfish mining strategy, the honest miners perform waste computations, the blocks they confirm will not be a part of the accepted blockchain. The selfish pool does not immediately publish the blocks they find (like it is required by the standard Bitcoin protocol). They selectively withhold block(s) and broadcast the chain they created at the right time.

Eyal and Sirer (2018) also claim that "above a certain threshold size, the revenue of the selfish pool rises super-linearly with the pool size in comparison with the honest revenue". Once the threshold has been breached, rational miners want to join the pool and miners already in the pool want to attract new participants to increase their revenue. At a certain point, the pool becomes so large, that it can control the blockchain and the selfish strategy becomes obsolete. A possible solution to this problem is: a miner should mine uniformly at random on all the competing branches (if he has any knowledge of them). However, Sapirshtein et al. (2017) find that "attackers with strictly less than 25% of the computational resources can still gain from selfish mining" even if this rule is implemented. On top of that, they show that people who can perform a profitable selfish miner attack, can perform a double spend attack at zero cost.

3.3.3   ICO

ICO, also known as initial coin offering, is a fund-raising operation in the cryptocurrency world. ICOs are not only used to launch new cryptocurrencies, it is also applied to inaugurate services and applications for companies, organizations and entrepreneurs. As a consequence, an ICO does not solely exist in the IT-space. There have been successful initial coin offerings in health care, energy, ... Fiat currency or other more established coins are invested in the ICO, in exchange the investors receive tokens specific to the ICO. There are two types of tokens, utility tokens and security tokens. Security tokens offer participation in governance and future earnings, while utility tokens promise access to the future services of the ICO. Many regulators believe that security tokens should be treated the same as securities in regular markets, but currently not much regulation is in place.

ICOs are important to business activities for four reasons. First, ICOs can serve as a mean to bypass the highly regulated capital-raising-process required for banks and venture capitalists. Secondly, using an ICO means that the organization can cut on the cost of capital raising, namely, an ICO cuts out the middle man (like crowdfunding platforms or banks). A third reason is that using an ICO to raise money often provides a built-in customer base and positive network effects (Giudici and Rossi-Lamastra 2018). Lastly, funders create a secondary market for their investment, the tokens can be sold to someone else.

However, ICOs are unfortunately also a hotspot for scams and fraud since there is no official prospectus in an ICO and the contributors are not protected. Moreover, the creators of an ICO often disclose limited information. The first ever ICO was organized by the DAO project and raised 150 million US Dollar. The DAO was developed on the Ethereum blockchain (as are many other ICOs because Ethereum is a programmable blockchain). In the end, hackers were able to steal over 60 million US Dollar due to a loophole in the code. This also presents another issue of ICOs, the ICO needs to be designed optimally such that hackers cannot benefit (Conley 2017; Yadav 2017).

An ICO can be thought of as a mixture between crowdfunding and an IPO (initial product offering). ICOs resemble crowdfunding in the lack of investor protection, the fact that there is no relevant track record, a limited set of information and no supervision by public authorities. However, crowdfunding collects solely fiat money through traditional payment channels. An ICO is decentralized and hardly regulated, therefore it differs fundamentally from IPOs. ICOs create digital tokens on a blockchain and these are distributed via a public ledger. On the other hand, IPOs distribute shareholdings through investment banks, who serve as underwriters. Every company can start an ICO, while IPOs are more exclusive. The idea behind IPOs and ICOs is the same, they both want to raise money for a company.

An ICO often outlines its plans in a white paper which provides details regarding the objective, the amount of money needed, how much of the virtual tokens are kept by the creators and the accepted currencies. If the ICO is unsuccessful, which means that the minimal amount of funds is not reached, the money of the investors is either returned or kept by the organizer.

Figure 3.6 provides an overview of the number of successful ICOs over the years. In 2014 and 2015 there were only 7 ICOs over the whole year, while from 2016 onward the number of ICOs per year grew exponentially, with already 460 successful ICOs in 2018. The fact that an ICO is a success relies on several factors, such as the availability of the code, bonus schemes and presale initiatives (preceding the ICO) (Giudici et al. 2018). These presale initiatives show that it is advantageous to test the market with a smaller token sale to see if the market is interested. The availability of the code (even if only parts of it are available) is important since

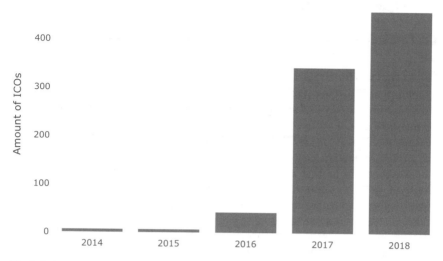

Fig. 3.6 Amount of successful ICOs as reported by CoinDesk (2019)

there is a general lack of information and the code allows contributors to asses the technical value and viability of the project (Giudici et al. 2018). Moreover, if source-code is open, then every programmer can aid in fixing bugs and testing.

As mentioned before, ICOs are hardly regulated, some central banks like the People's Bank of China even banned this kind of money-raising (Huillet 2018; Emem 2018). This particular announcement plummeted the price of Bitcoin and Ethereum. Many social media applications, like Facebook and Twitter, have banned advertisement of ICOs (Leathern 2018). This is a logical reaction to the number of scam ICOs. In 2017 alone, over 80% of all the ICOs have been reported to be fraudulent; however, studies showed that less than 30% of all the money raised using ICOs was invested in these scam projects (Alexandre 2018). ICOs are not entirely unregulated, the SEC implemented several rules for ICOs. For example, if ICOs uphold some characteristics, they can be ruled as securities offerings, in this case the ICO falls under SEC jurisdiction and the federal securities law can be applied (SEC 2019). The European Union and more specific the ESAs (European Supervisory Authority) are drafting up legislation for ICOs in 2019. Further legislation on ICOs could attract more established investors and could be promising for the crypto-market. However, the "wrong" legislation could also have devastating effects for the development and price of cryptocurrencies and their market.

3.4 Operational Risk

3.4.1 Wallet

Crypto-wallets (Blockgeeks 2019; Cryptocurrency Facts 2017) actually do not resemble their real-life counterparts. Once a coin is mined, the users do not actually keep the coin in a physical safe place, since the cryptocurrency system only stores the transactions and not the coins. In other words, once a miner mines a coin, a new transaction of the form: "BTC address A has mined X amount of BTC" is created. The amount of money someone owns is the result of all the transactions that have been written on the blockchain for their own address (public key). In general, it is the history of all the transactions that reveals the owner of the coins. For this particular reason, regular wallets are completely different from cryptocurrency wallets; crypto-wallets protect the private key, used to sign transactions off. It hence actually is not really storing the funds, it merely stores the public and private key of the user and interacts with various blockchains to allow the user to check their balance, send money and do other operations. Cryptocurrencies are assigned to public addresses, in other words, users have ownership over the coins if the private key of their wallet matches the public key behind the coins. The wallet basically is a piece of software designed to interact with blockchains.

There exists several types of wallets. In general, a wallet is able to interact with different blockchains and is capable to store several crypto-specific public addresses. One of the most secure wallets is the hardware wallet. These wallets go online to perform transactions, but the remaining time they are offline which significantly decreases the probability of being hacked. Online wallets operate via the cloud, therefore the private keys are also stored online and thus are more prone to hacks. Paper wallets are another kind of wallet, they print a QR-code of both the public and private key. This option allows users to completely avoid storing digital data regarding their cryptocurrencies. One thing that holds for every wallet is that if users lose their private key, they also lose access to their funds. Moreover, once money is erroneously transferred to another account, it is not possible to get it back.

Since wallets are a piece of software that is constantly improved, it is also subject to regular maintenance and security updates, during these updates users cannot access their funds. Hence, if a transaction is time-sensitive, these updates can cause some disruptions.

3.4.2 Exchanges

Cryptocurrency systems are used to exchange real goods and can have a bidirectional flow with real currencies. Users convert cryptocurrencies to fiat money or another cryptocurrency through exchanges. There exists two types of exchanges, the first kind allows its users to exchange regular currency or cryptocurrency for

Table 3.1 Most important exchanges according to trading volume on 10 September 2019

#	Exchange	24 h Volume (in million)	# Coins
1	Binance	743.922	167
2	Bitfinex	83.336	34
3	Gate.io	34.008	191
4	Bittrex	15.562	249
5	Poloniex	11.998	56
6	Coinbase Pro	125.301	21
7	Huobi Global	496.294	224
8	Kraken	105.184	22
9	Upbit	74.372	203
10	Bitstamp	65.179	5

another cryptocurrency or visa versa. The second kind is a derivative exchange where users can take positions on the future price of cryptocurrencies. The first kind of cryptocurrency exchanges provide a digital platform to buy and sell cryptocurrencies at spot price. This means that the prices are immediately agreed upon. Cryptocurrency exchanges offer their services against a small conversion or transaction fee. Table 3.1 gives a ranking of cryptocurrency exchanges according to 24 h trading volume on the 10th of September 2019. The exchanges who have a high trading volume are Binance and Bitfinex.

The cryptocurrency derivative exchange does not trade in cryptocurrencies itself but rather in financial contracts, where cryptocurrencies are the underlying. These contracts can be used to hedge against the highly volatile nature of cryptocurrencies or for a pure speculative basis. The first cash settled future contracts on Bitcoin/Dollar were traded on the Chicago Board Options Exchange (CBOE) in December 2017, not long after the CME group followed the example. LedgerX or Derebit are examples of other derivative exchanges for cryptocurrencies.

Exchanges can create operational risk in the sense that these systems need maintenance. The major established exchanges have the custom to announce maintenance in due time, however, there are many instances recorded of some exchanges going completely unannounced partially or even completely offline for several days before coming live again. Maintenance can also affect just some parts of the exchange like the depositing or withdrawal of particular cryptocurrencies. Hence, inter-exchange arbitrage trading can be seriously affected by such incidents. If certain currencies trade at different prices at different exchanges, one can profit of this by inter-exchange arbitrage trading. One is then buying one cryptocurrency at one exchange, then is transferring this to another exchange, to sell it there ideally at a higher price. Of course such a transaction is subject to a certain time lag, essentially the transfer time. Whether it eventually is profitable, depends on the confirmation time of transactions in the blockchain (transferring the cryptocurrencies from one exchange address to another exchange address). If in the meantime some maintenance takes place, obviously the transfer, handled by the exchange, can be significantly delayed and hence the arbitrage could disappear before the transfer is finalised and the sell order is given.

Furthermore, cryptocurrency exchanges are not subject to the regulations of regular exchanges, they even operate with lower safeguard than traditional markets. Moreover, some platforms conduct overlapping lines of business that present serious conflicts of interest, including trading for their own account on their own venues. Another issue is that exchange platforms lack real-time and historical market surveillance which should enable them to detect suspicious behaviour in the market, to track money-laundering for example. The biggest risk is perhaps the risk of losing trust in the exchange and as a consequence all the users ask for their money back at the same time. Part of the money could have been invested elsewhere and thus not all the requests can be granted at the same time. There is hence a potential risk of a exchange run—similar like one has a bank-run.

Finally, exchanges are also vulnerable of hacking. If a hacker can take control of an exchange, it essentially can transfer all it users' currencies to its own anonymous address. For an exchange it is hence very important to protect the exchange against such hacks and to ensure good cybersecurity to safeguard the trust of the users. Therefore, the exchange user is exposed to clear counterparty risks. However, it happens that cryptocurrency exchanges still get hacked. On 11 June 2019 a Japanese exchange, called Bitpoint, experienced a major hack, during which the exchange lost 32 million USD worth of cryptocurrencies (Hickey 2019).

3.4.3 Other Operational Risks

We will now briefly mention two other operational risks, which investors need to keep in mind. First of all, cryptocurrencies have been used for money laundering practices in the past and criminals still believe cryptocurrencies offer a good alternative to the more traditional money launder practices. The money of the criminals is converted in different cryptocurrencies and then transferred to the digital wallets of the criminal organization often going through mixing processes (Cryptalker 2019; Malwa 2018) which try to obscure the origin of the coins. In July 2018 Europol successfully dismantled two criminal organizations that were involved in these kind of practices in Spain (Europol 2018), in this specific case the criminals used Bitcoin to launder the money.

Another operational risk of cryptocurrencies is the large amount of energy needed to keep the system running. It has been reported that the increasing demand in energy has led to numerous blackouts in Venezuela alone (Holthaus 2017). Iceland has similar fears since the country has attracted many crypto-companies (Noack 2018). The problem is most severe in Proof-of-Work cryptocurrencies, since mining the next block takes up a lot of energy.

References

Alexandre A (2018) New study says 80 percent of ICOs conducted in 2017 were scams. https://cointelegraph.com/news/new-study-says-80-percent-of-icos-conducted-in-2017-were-scams

Blockgeeks (2019) Cryptocurrency wallet guide

Brito J, Shadab HB, Castillo A (2014) Bitcoin financial regulation: securities, derivatives, prediction markets, and gambling. Columb Sci Technol Law Rev 16:221

CoinDesk (2019). https://www.coindesk.com

Coinmarketcap (2019) Market capitalisation of cryptocurrencies. https://coinmarketcap.com/

Coinranking (2019) Coinmarketcap. https://coinranking.com/

Conley JP (2017) Blockchain and the economics of crypto-tokens and initial coin offerings

Copeland T (2019) Cryptocurrencies protect 51 attacks. https://decryptmedia.com/4408/cryptocurrencies-protect-51-attacks

Cryptalker (2019) 10 best Bitcoin tumbler (mixer) services: review 2019. https://cryptalker.com/best-bitcoin-tumbler/

Cryptocurrency Facts (2017) What is a cryptocurrency wallet? https://cryptocurrencyfacts.com/what-is-a-cryptocurrency-wallet/

Emem M (2018) Illegal financial activity PBoC deputy governor warns against STOs in China. https://www.ccn.com/illegal-financial-activity-pboc-deputy-governor-warns-against-stos-in-china

European Commission (2018) Fintech action plan: for a more competitive and innovative European financial sector. Technical report

Europol (2018) Two criminal groups dismantled for laundering 2.5 through smurfing and cryptocurrencies. https://www.europol.europa.eu/newsroom/news/two-criminal-groups-dismantled-for-laundering-eur-25-million-through-smurfing-and-cryptocurrencies2

Eyal I, Sirer E (2018) Majority is not enough: Bitcoin mining is vulnerable. Commun ACM 61(7):95–102

Frankenfield J (2019) 51% attack. https://www.investopedia.com/terms/1/51-attack.asp

Giudici G, Rossi-Lamastra C (2018) Crowdfunding of SMEs and startups: when open investing follows open innovation. Res Open Innov SMEs 377–396

Giudici G, Adhami S, Martinazzi S (2018) Why do businesses go crypto? An empirical analysis of initial coin offerings. J Econ Bus 100:64–75

Goffard P-O (2018) Fraud risk assessment within blockchain transactions. IDEAS Working Paper Series from RePEc

Goyal S (2019) 51% attack explained: the attack on a blockchain. https://www.fxempire.com/education/article/51-attack-explained-the-attack-on-a-blockchain-513887

Hickey S (2019) $32m stolen from Tokyo cryptocurrency exchange in latest hack. https://www.theguardian.com/technology/2019/jul/12/tokyo-cryptocurrency-exchange-hack-bitpoint-bitcoin

Holmes S (2019) China says new digital currency will be similar to facebook's libra. Reuters

Holthaus E (2017) Bitcoin could cost us our clean-energy future. https://grist.org/article/bitcoin-could-cost-us-our-clean-energy-future/

Huillet M (2018). China's central bank warns investors of ICO, crypto risks. https://cointelegraph.com/news/chinas-central-bank-warns-investors-of-ico-crypto-risks

Komodo (2018) Daniel. Security: delayed proof-of-work. https://komodoplatform.com/security-delayed-proof-of-work-dpow/

Kroll JA, Davey IC, Felten EW (2013) The economics of Bitcoin mining, or Bitcoin in the presence of adversaries. In: The twelfth workshop on the economics of information security (WEIS 2013), pp 11–21

Leathern R (2018) New ads policy: improving integrity and security of financial product and services ads. https://www.facebook.com/business/news/new-ads-policy-improving-integrity-and-security-of-financial-product-and-services-ads

Malwa S (2018) $1.2 billion in cryptocurrency laundered through Bitcoin tumblers, privacy coins. https://www.ccn.com/1-2-billion-in-cryptocurrency-laundered-through-bitcoin-tumblers-privacy-coins

Mersch Y (2018) Virtual or virtueless? The evolution of money in the digital age. Lecture by Yves Mersch, Member of the Executive Board of the ECB, Official Monetary and Financial Institutions Forum

Noack R (2018) Cryptocurrency mining in Iceland is using so much energy, the electricity may run out. https://www.washingtonpost.com/news/worldviews/wp/2018/02/13/cryptocurrency-mining-in-iceland-is-using-so-much-energy-the-electricity-may-run-out/?noredirect=on&utm-term=.23a7a64db9e2

Osato A-N (2018) 51 percent attack: Hackers steal 18 million Bitcoin gold (BTG) tokens. https://bitcoinist.com/51-percent-attack-hackers-steals-18-million-bitcoin-gold-btg-tokens/

Pieters G, Vivanco S (2017) Financial regulations and price inconsistencies across Bitcoin markets. Inf Econ Policy 39(C):1–14

Rosenfeld M (2014) Analysis of hashrate-based double spending. arXiv.org

S J (2018) What is a 51% attack or double spend attack. https://medium.com/coinmonks/what-is-a-51-attack-or-double-spend-attack-aa108db63474

Sapirshtein A, Sompolinsky Y, Zohar A (2017) Optimal selfish mining strategies in Bitcoin. In: Grossklags J, Preneel B (eds) Financial cryptography and data security. Springer, Berlin, pp 515–532

Schrijvers O, Bonneau J, Boneh D, Roughgarden T (2017) Incentive compatibility of Bitcoin mining pool reward functions. In: Lecture notes in computer science (including subseries lecture notes in artificial intelligence and lecture notes in bioinformatics). Springer, Berlin, vol 9603, pp 477–498

SEC (2019) ICO. https://www.sec.gov/ICO

Yadav M (2017) Exploring signals for investing in an initial coin offering (ICO). SSRN Electronic J

Chapter 4
Quantitative Risks

Abstract The cryptocurrency asset prices are a rich source of information on cryptocurrency behavior. Therefore, we will investigate the historical time series of the price process of Bitcoin and other cryptocurrencies to pinpoint particular characteristics. We will start with a thorough descriptive analysis of the log returns of the most well-known cryptocurrencies. Afterwards, we will investigate the inter-correlation between cryptocurrencies and also the correlation between cryptocurrencies and other asset classes. To end this chapter, we will conduct an analysis of some quantitative risks by zooming in on the value at risk of cryptocurrencies.

Keywords Correlation · Peaked distribution with large positive outliers · Fat tails · Stationary time series · Volatility clustering · Highly intercorrelated cryptocurrency market · Flash crash · Pump and dump · Cauchy distributed · T-distributed · ARMA(2 · 2)-GARCH(1 · 3) · Long run dependency · Value at risk · Market efficiency

4.1 Descriptive Analysis of Returns

Next, we perform some time-series analysis on a variety of cryptocurrencies and further also investigate their correlation with other more traditional assets. One must be cautious with cryptocurrency data since it varies considerable (Alexander and Dakos 2019). We typically will work with the so-called log-returns over a single time unit (in most cases 1 day):

$$r_t = \log\left(\frac{P_t}{P_{t-1}}\right) \tag{4.1}$$

where t denotes the time and P_t is the asset price at a certain time t. We will use the daily historical prices of Ripple, Monero, Bitcoin and Ethereum to sketch an overview of the distributional and volatile behaviour of cryptocurrencies. Note that

Table 4.1 Descriptive analysis of daily log returns

	Min.	Median	Mean	Max.	Yearly std.	Kurt.	Skew.
BTC	−0.225	0.003	0.002 (0.117)	0.287	0.645	5.869	−0.218
ETH	−0.319	−0.000	0.004 (0.250)	0.312	1.027	3.466	0.058
XMR	−0.285	0.001	0.004 (0.433)	0.625	1.166	9.102	1.116
XMR*	−0.285	0.001	0.003 (0.143)	0.528	1.096	5.923	0.655
XRP	−0.496	−0.003	0.003 (0.179)	0.881	1.194	26.023	2.562
XRP*	−0.496	−0.003	0.002 (0.279)	0.593	1.085	14.013	1.471
XAU	−0.018	0.000	0.000 (0.410)	0.024	0.097	0.620	0.068
BCOM	−0.027	0.000	−0.000 (0.612)	0.027	0.104	1.492	−0.316
S&P500	−0.042	0.001	0.000 (0.286)	0.048	0.130	6.122	−0.621
EURUSD	−0.024	0.000	−0.000 (0.819)	0.030	0.098	2.311	0.101

The value between the parenthesis represents the *p*-value of the hypothesis of zero average returns. The starred (*) tickers represent the returns without the maximal return

cryptocurrency exchanges trade 24/7. We will mainly focus on the historical closing prices starting from the first of January 2016 onward.

Table 4.1 gives a descriptive analysis of the log returns of the different cryptocurrencies and regular assets. Note that in comparison to other more traditional assets, the difference between the maximal and minimal value of the log returns of cryptocurrencies is high, as is the yearly standard deviation (std.). The median is in each case for cryptocurrencies smaller than the mean, this indicates the presence of positive outliers. The value in between parenthesis represents the *p*-value of testing the null-hypothesis of zero average returns. For all the returns under examination, the null-hypothesis cannot be rejected.

All the log returns of cryptocurrencies also exhibit a higher kurtosis (kurt.) than the normal distribution, this means that the distribution is likely to have fat tails and will exhibit more peakedness. The kurtosis of the cryptocurrencies is comparable to those of emerging government bonds. The skewness (skew.) of the log returns is positive except for Bitcoin. This positive skewness is not present in typical equity markets. We also observe that commodities (gold and the commodity index) and the Euro Dollar exchange rate have a kurtosis lower than 3, the kurtosis of a normal distribution, while the S&P500 stock index has a higher kurtosis. The commodity index and the S&P500 stock index both have a negative skewness, while gold and the exchange rate have a positive skewness.

A comparison among the log returns of the cryptocurrencies shows that Ripple has simultaneously the smallest and the largest log returns on his name. Moreover, it has the highest yearly standard deviation among all the cryptocurrencies. From a historical time-series perspective, Ripple is hence the most volatile of all the cryptocurrencies under consideration. When we delete the maximal return, as depicted with the starred ticker, we can still conclude the same remarks. Bitcoin is the least volatile of all the cryptocurrencies which are under examination here. Ripple exhibits the highest kurtosis and skewness, which means that Ripple has heavier tails and more large outliers than the remaining cryptocurrencies. The

Table 4.2 Result of different stationarity tests

	ADF		KPSS		PP	
	Test statistic	P-value	Test statistic	P-value	Test statistic	P-value
BTC	−33.861	0.000	0.340	0.105	−31.120	0.000
ETH	−17.836	0.000	0.605	0.022	−34.790	0.000
XMR	−11.305	0.000	0.476	0.046	−36.319	0.000
XRP	−7.165	0.000	0.147	0.399	−35.356	0.000
XAU	−25.510	0.000	0.195	0.363	−29.201	0.000
BCOM	−25.395	0.000	0.172	0.330	−29.981	0.000
S&P500	−8.984	0.000	0.115	0.516	−23.842	0.000
EURUSD	−42.172	0.000	0.142	0.414	−42.417	0.000

median of the log returns of all the cryptocurrencies are similar and very close to zero.

The log returns of all the investigated assets from 01/01/2016 until 25/02/2019 are probably stationary according to the Augmented Dickey Fuller (ADF) and Phillips Perron (PP) test. The tested null-hypothesis is:

H_0 : The time series is non-stationary (not necessarily with a trend).

Stationary timeseries are timeseries whose unconditional joint probability remains the same over time, the weak stationarity only requires the (co)variance and mean to be constant over time. On the other hand, non-stationary behaviour can have cycles, random walks and trends or a combination of these three. Table 4.2 shows that the null-hypothesis is rejected at 5% significance level because the p-value for the ADF- and PP-test is smaller than 0.05, which means that the process is stationary. The null-hypothesis test for the Kwiatkowski Phillips Schmidt and Shin (KPSS) test is

H_0 : The time series is weakly stationary.

The null-hypothesis is rejected for Ethereum and Monero at a 5% significance level because the corresponding value in Table 4.2 is smaller than 0.05. For all the other log returns the weakly stationary hypothesis is not rejected.

Figure 4.1 shows one important remark, the log returns of cryptocurrencies do exhibit periods where high returns cluster together and periods where low returns reside, this could point to volatility clustering in the log returns. Another observation is that half of the time the returns are positive and half of the time the returns are negative. The daily log returns of cryptocurrencies seem to be mean-reverting to zero and Fig. 4.2 provides some further examination of the log returns over time. The cumulative returns over time from mid 2015 until February 2019 is shown. The cryptocurrencies have had an upward movement until the end of 2017. From 2018 onward mainly negative returns were realized. In comparison to the traditional assets, cryptocurrencies have had a much higher cumulative return.

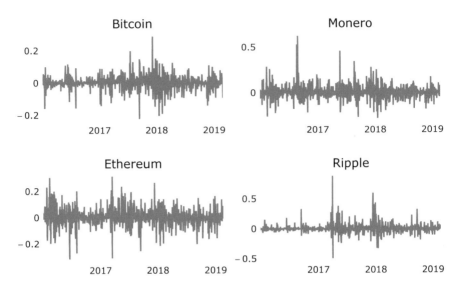

Fig. 4.1 Log returns of the different cryptocurrencies

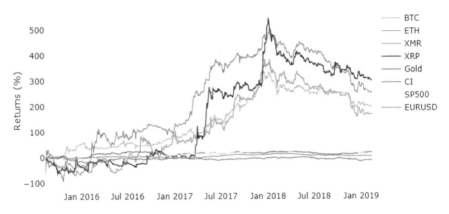

Fig. 4.2 Cumulative log returns

Cryptocurrencies often exhibit behaviour where large returns happen due to crashes in the price. In November 2018, the price of Bitcoin began to drop. The drop started with a crash of nearly 10% between the 13th of November and the 14th of November after nearly a month of having a stable price around 6200 USD, see Fig. 4.3a. Moreover, the whole market dropped from a stable market price of 209 billion USD to 187 billion USD, which is a drop of 10.53% effectively. The mid November drop is the start of the price decline of Bitcoin, the level of 13 November 2018 will not be reached again in the near future. Between 24 February and 25 February the price of Bitcoin again suffered from a crash, as Fig. 4.3b clearly

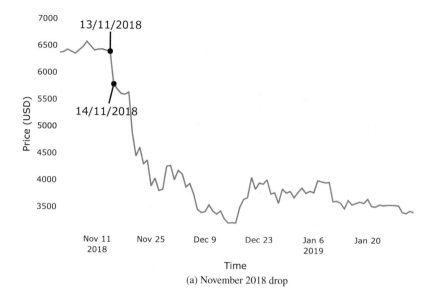

(a) November 2018 drop

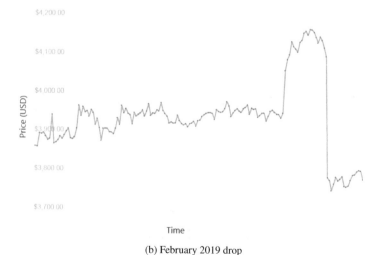

(b) February 2019 drop

Fig. 4.3 Price drops in Bitcoin (**a**) represents a price drop from 13/11/2018 until 14/11/2018 of nearly 10%. (**b**) Depicts a drop of 9.7% during the month February of that same year

shows. This time the price of Bitcoin dropped 9.7% over the course of only hours. Similar behaviour as in Fig. 4.3b is found in Ethereum, Litecoin and many other cryptocurrencies. This could be a consequence of the leader (Bitcoin) follower behaviour in cryptocurrencies.

4.2   Correlation and Diversification

4.2.1   Correlation Between Several Cryptocurrencies

Figure 4.4a, b and c represent heatmaps of the Pearson correlation between the daily log returns of several cryptocurrencies from 2016 until 2018. Bitcoin (BTC) and Litecoin (LTC) exhibit a high correlation in the year 2016, which is no surprise since both currencies have similar characteristics and prospects. The strong positive dependence between these two cryptocurrencies continues to exists for following years. As discussed, Ripple (XRP) and Stellar (XLM) also have similar characteristics, since they both act in the remittance field. The correlation between them also support this claim through the years. Note that EOS did not exist in 2016.

Note that the correlation between several cryptocurrencies has significantly increased over the last 3 years. In other words, not only currencies with similar characteristics have high correlation, but the whole market seems to move in the same direction. One particular reason for the cryptomarket to be so correlated

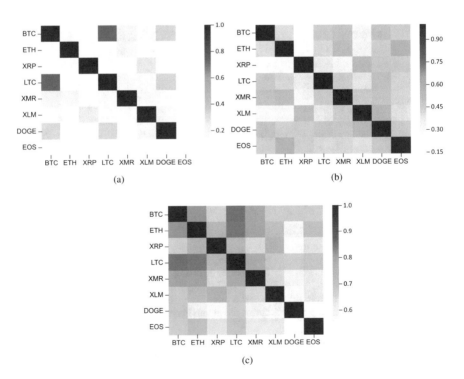

Fig. 4.4 Correlation heatmaps between different cryptocurrencies through the years. (**a**) 2016. (**b**) 2017. (**c**) 2018

is because many cryptocurrencies are bought using Bitcoin or Ether. Hence, the dominant position of Bitcoin or Ethereum in the market is fortified.

4.2.2 *Correlation Between Cryptocurrencies and Other Stocks*

Next, we investigate the correlation of cryptocurrencies with various other types of assets: commodities, equity and currencies. The commodity class will be represented by gold, the security class by S&P500 stock index and to model the currency behaviour we use the EUR/USD exchange rate.

First, we investigate the correlation between the log returns of Bitcoin and the S&P500 stock index. Figure 4.5a shows the rolling Pearson correlation over a window of a predefined number of days between the log returns of both assets. The assets appear to become less correlated if the number of data points to construct it becomes larger. Moreover, the correlation, when taking into account 180 data points per window, is nearly zero, but positive. Besides, short term correlations fluctuate around zero. The correlation between Bitcoin and the S&P500 based on a window of 180 days never exceeds 30%, which indicates that Bitcoin is not very correlated to it.

The correlation between Bitcoin and the price of gold in USD is shown in Fig. 4.5b. A rolling window of respectively 10, 60 and 180 days is used to find the correlation over time. The 10-day correlation fluctuates around zero, like the correlation between S&P500 index and Bitcoin. The 60-day correlation, in this case is mainly negative, this means that the price movements in gold and Bitcoin act in opposite directions. Moreover, the 180-correlation is very close to zero but negative, which means that the price movements of gold and Bitcoin are more or less uncorrelated or move in opposite directions.

The Pearson correlation between the log returns of EUR/USD exchange rate and Bitcoin is plotted in Fig. 4.5c. The 10-day correlation fluctuates around zero, however, longer time-periods lead to a more negative correlation. The 180-day correlation is even persistently below zero. Note that the 60-day correlation goes up in the end, hence it is likely that the 180-day correlation will follow this pattern.

The one year correlation between Bitcoin and other major asset classes (bonds, oil, US real estate and emerging market currencies) stays within the boundaries of being a differentiated risk reducer (see Fig. 4.5d) (Burniske and White 2017) because the correlation remains under 0.3 in absolute value. The asset classes are quantified by

- Bond: Bloomberg Barclays US aggregate bond index (LBUSTRUU)
- Oil: crude oil futures (CL1 Comdty)
- Emerging markets currencies: J.P. Morgans emerging markets currency index (FXJPEMCS Index)
- Real estate: Vanguards US real estate ETF (VNQ).

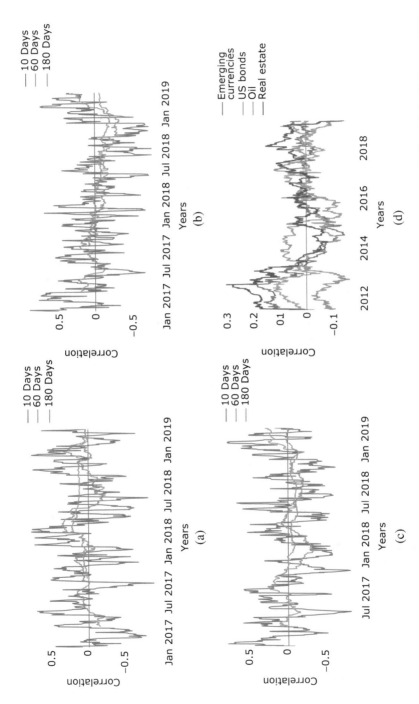

Fig. 4.5 The rolling correlation with window sizes of 10, 60 and 180 days. (**a**) Between BTC and S&P500. (**b**) Between BTC and Gold. (**c**) Between BTC and Euro Dollar exchange rate. (**d**) Between Bitcoin and other assets

Moreover, the average correlation with these major asset classes is fluctuating around zero, which means that Bitcoin is not heavily dependent on other assets.

4.3   Flash Crash

A flash crash (Kenton 2018) is a sudden steep decline in price of the underlying asset. Many people think flash crashes are a part of artificial market manipulation. There are several reasons why flash crashes can happen; human error, fraud, high frequency trading and computer/software glitches are all possible triggers. High frequency trading is an automated trading system, where computers are able to detect changing market conditions and make trades accordingly. These computers place enormous orders at light speed and as a result they can enlarge the negative price movements. Each crash is unique and it is difficult to pinpoint the exact cause. Many cryptocurrencies exchanges are open via an API system to all kinds of trading bots. Some bots try to play the bid ask spread, others do automated trading on the basis of technical analysis or other kinds of algorithms. The presence of such bots, often written by non-experienced participants increase the risk of flash-crashes.

In May 2019 the BTC-Canadian Dollar (CAD) price experienced a flash crash of 99.1% on the exchange Kraken (SFox 2019), see Fig. 4.6. The price for one Bitcoin dropped from 11 800 CAD to a low of 101.2 CAD. Only a few minutes after the flash crash, the price again reached its normal level. One of the possible reasons of the crash is the limited liquidity on Kraken to exchange Bitcoin for Canadian Dollar. Some traders were very lucky to obtain Bitcoins at the low price. However, this also means that the sellers of those particular Bitcoins made a huge loss. Stop-loss-orders can be the reason why there were Bitcoins available in the market at such a low price.

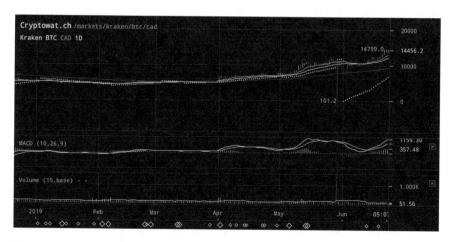

Fig. 4.6   Flash crash of BTC/CAD from cryptowat.ch

A stop-loss-order is an order which is send to the market once a particular price-level is breached. If the level is breached, the stop-loss-order becomes an order. In this case, the order became a sell order at the market price during the crash.

4.4   Pump and Dump

Market manipulation is a crime in regulated markets, however many crypto-exchanges are unregulated and hence are under no real external supervision. This makes them vulnerable for all kinds of manipulation. Pump and dump is a form of such market abuse, in the regulated markets this scheme is illegal. However, since crypto-markets are mostly unregulated, it occurs much more frequently there. Some players drive the price of a stock they hold up (pump), by creating false interest in it. Once the price has risen sufficiently high, they sell their stocks on the market at an higher price than the original (true) market price (dump). The objective of a pump and dump scheme practitioner is distorting supply and demand in their favour. In general, pump and dump actions tend to work better on small illiquid stocks, because then a sharp increase in trade volume can send the price up drastically.

In cryptocurrencies there exists organized groups who perform pump and dump schemes. The groups post 'insider tips' regarding cryptocurrencies, which promise quick gains and wealth. The participants in a pump and dump scheme are told when and where to buy the specified coin via messages. Usually a pump and dump scheme drives the trading volume and price up simultaneously.

Figure 4.7 shows two obvious pump and dump operations, one on the 7th of February 2019 and one on the 18th of February 2019. From this figure, one can clearly see that the volume and price, on these days, has increased significantly. After the initial wave of buying, the participating investors need to act quickly to sell the asset, while at the same time they encourage non-suspecting investors to buy the asset. The pump and dump group tries to convince the non-suspecting investors to buy the asset by promoting the asset and as such try to find support for the upward price move. The wave of selling the assets, by the participants in the pump and dump, decreases the price often even more than its original level.

Telegram channels advertise pump and dumps schemes nearly daily, sometimes even multiple times a day. Since the exchanges of cryptocurrencies are hardly regulated or even unregulated today.

4.5   Distributional Behaviour

The timeseries of the log returns of Bitcoin do not follow a white noise process. This can be seen by comparing Fig. 4.8a, which represents the log returns of Bitcoin over time, and Fig. 4.8b, which depicts a typical white noise process. The log returns of Bitcoin exhibit a more volatile pattern with so-called volatility-clusters (periods of

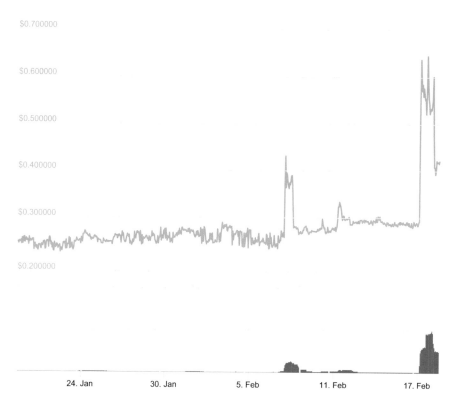

Fig. 4.7 The upper panel depicts the price history of Vertcoin and the lower panel the trading volume of Vertcoin

high volatility). There are indications that the log returns are stationary, according to the tests in Table 4.2, but the price movement in different time periods vary quite a lot. As mentioned before, the log returns of cryptocurrencies exhibit volatility clustering, it is important to accurately model the time variability of the volatility.

Figure 4.9 shows the quantile plots of the different cryptocurrencies their log returns. In this case, the quantiles of the Bitcoin, Ethereum, Ripple and Monero their log returns are compared to quantiles of the normal distribution. One can immediately see that the log returns have far heavier tails than the normal distribution suggests. Hence, the distributions of cryptocurrencies has to be chosen among the distributions which allow for fat tails. The most commonly used distributions with heavy tails are:

- Pareto distribution,
- Burr distribution,
- Cauchy distribution,
- Lognormal distribution,

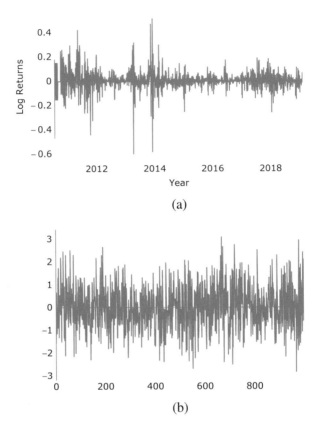

Fig. 4.8 Difference in log returns. (**a**) Bitcoin. (**b**) White noise

- Weibull distribution,
- Frechet distribution,
- T distribution.

Next, these distributions are used to find an appropriate distribution for the log returns of cryptocurrencies. If $x_1, \ldots, x_n$ are independent observations of X, then the maximum-likelihood parameters of each distribution are the values maximizing the likelihood

$$L(\mathbf{\Theta}) = \prod_{i=1}^{n} f(x_i; \mathbf{\Theta})$$

where $\mathbf{\Theta} = (\theta_1, \ldots, \theta_s)$ is a vector of parameters specifying $f(\cdot)$. In order to check which distribution with the optimal parameters implemented gives the best fit, the Kolmogorov-Smirnov (KS) test statistic is used. KS-test performs a test to see if

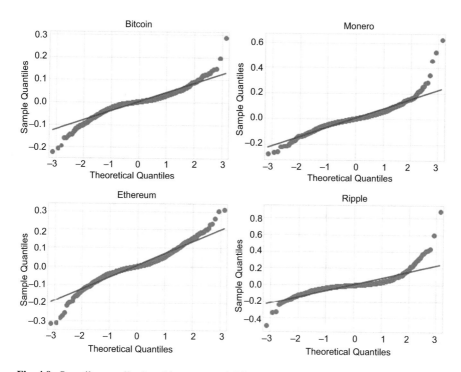

Fig. 4.9 Quantile-quantile plot of log returns of different cryptocurrencies

the distribution $G(x)$ of an observed random variable matches a given distribution $F(x)$. The null-hypothesis states that the two distributions are equal.

Using these optimal parameters, one can compare all of the distributions of the specified dataset. If more than one distribution offers a good fit according to the Kolmogorov-Smirnov test statistic, the distribution which offers simultaneously the lowest AIC and BIC value is chosen. AIC/BIC selects the best fit according to maximum likelihood and at the same time punishes for the number of parameters, as shown in Eqs. (4.2).

$$AIC = 2k - 2\ln\left(L_{max}\right)$$
$$BIC = \ln(n)k - 2\ln\left(L_{max}\right)$$

(4.2)

In Eq. (4.2) n denotes the number of observations, k is the number of estimated parameters and L_{max} represents the maximum of likelihood function.

Table 4.3 represents the outcome of the KS-test statistic for all of the data samples. The Kolmogorov-Smirnov test tests if the data comes from a predetermined distribution. In other words, the null-hypothesis is

H_0: the data comes from the specified distribution.

Table 4.3 Kolmogorov-Smirnov test on the log returns of different cryptocurrencies

	Cauchy distribution		T distribution	
	Test statistic	P-value	Test statistic	P-value
BTC	0.032	0.354	0.038	0.187
ETH	0.041	0.032	0.039	0.055
XMR	0.043	0.009	0.023	0.438
XRP	0.030	0.234	0.030	0.252

For the log returns of Bitcoin, the null-hypothesis for the t and Cauchy distribution cannot be rejected, since the p-value is larger than the significance level 0.05. Therefore, both distributions offer a good fit. However, the t distribution has a lower AIC and BIC value, therefore this distribution seems currently the most appropriate distribution in terms of fitting the historical time series. Figure 4.10a displays all of the fitted distributions to the empirical data of the log returns of Bitcoin. The distribution which fits the log returns of Bitcoin the best, according to AIC/BIC and the KS-test statistic, is the t distribution with location parameter 0, scale parameter 0.02 and the degrees of freedom are equal to 1.72. Moreover, we find similar results by doing the same analysis for Ethereum, Monero and Ripple their log returns. All the other cryptocurrencies also choose the t distribution as the best fitting distribution according to AIC/BIC and KS-test. These results can be seen visually in Fig. 4.11.

4.6   Volatile Behaviour

Cryptocurrencies are highly volatile assets. Fiat currency, like Yen, Dollar or Euro, do not fluctuate much, while cryptocurrencies have seen their fair share of severe price movements. The highly volatile nature does not allow cryptocurrencies to accurately convey relative prices of goods and services in the economy and leads to uncertainty to its holders regarding its value. In the next chapter, we will have a closer look at the volatile behaviour of Bitcoin.

The analysis is restricted to a gold price index ("XAU Curncy"), a commodity index ("BCOM") and the Euro Dollar exchange rate ("EURUSD") for the non-crypto assets. Figure 4.12 shows the yearly volatility in percentage of different 'securities' measured by standard deviation of the log returns using a rolling window of 10 days, we also include the VIX for comparison. VIX stands for the Chicago Board Options Exchange volatility index, which measures the implied volatility of S&P500 index options. The daily volatility at time t is calculated as:

$$\sigma_t = \sqrt{\frac{\sum_{s=1}^{w}(r_s - \bar{r})^2}{w}}$$

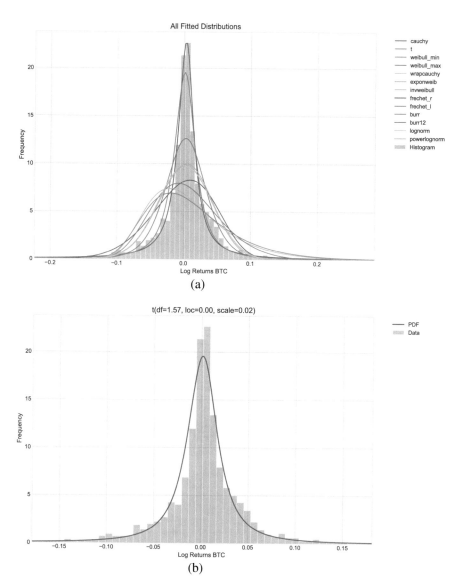

Fig. 4.10 Distribution of Bitcoin's log returns. (**a**) Tests. (**b**) Final

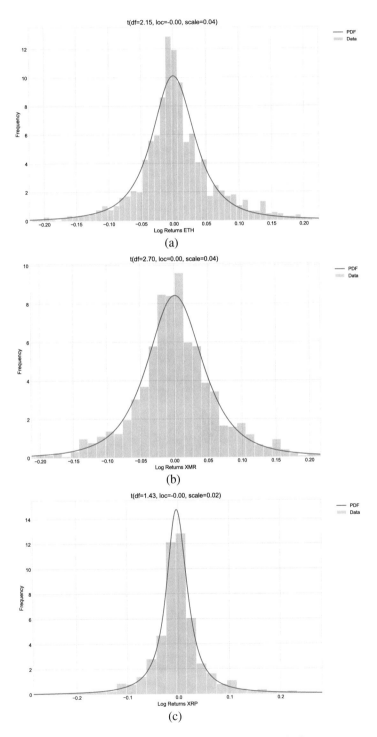

Fig. 4.11 Distribution of log returns. (**a**) Ethereum. (**b**) Monero. (**c**) Ripple

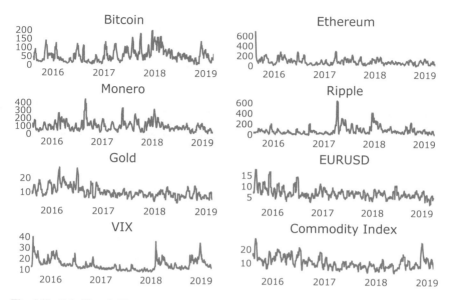

Fig. 4.12 Volatility of different 'securities' in percentage, quantified by yearly standard deviation of the log returns using a rolling window of 10 days

where $\bar{r}$ is the average return over the window (w), the yearly standard deviation is obtained by multiplying the daily standard deviation with $\sqrt{365}$. The data span a period from 10/08/2015 until 14/02/2019. One thing that is immediately clear from Fig. 4.12 is that cryptocurrencies are persistently more volatile than regular securities, some days the yearly standard deviation is ten times higher.

The beginning of 2018 also seems to be the start of a new period volatility-wise for cryptocurrencies. This is around the same time that futures on Bitcoin were traded on exchanges, people now take short-positions on the Bitcoin market more easily and this apparently has calmed down the high volatility.

4.6.1   ARMA-GARCH Model

In this section, we will fit an ARMA-GARCH model to the daily log returns of Bitcoin from 01/01/2016 until 17/11/2018, this leaves a total of 1051 observations for the fitting. The data from 18/11/2018 until 25/02/2019 will be used as test data to check the goodness of fit of the ARMA-GARCH model. As mentioned before, the log returns of cryptocurrencies exhibit excess kurtosis and fat tails, these are typical evidences of heteroskedastic effects such as volatility clustering. Moreover, Fig. 4.13 represents the squared log returns of Bitcoin, which fluctuate around a constant level. However, they also exhibit periods where large and small changes of log returns cluster together, this can indicate volatility clustering.

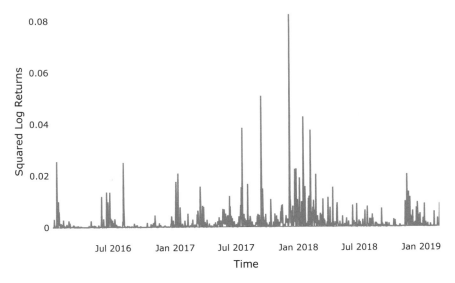

Fig. 4.13 Squared log returns of Bitcoin

We also test for long-run-dependency (LRD) of the returns and the volatility. We will calculate the LRD using the Hurst parameter (Bacon 2008). Figure 4.14a and b represent the rolling monthly Hurst parameter of respectively the returns and the volatility of Bitcoin. From these Figures, it can be concluded that the returns and the volatility of the returns have a Hurst exponent smaller than 0.5, this means that the these time-series are anti-persistent. In other words, they fluctuate violently but are mean-reverting.

To further investigate this behaviour, the sample partial and regular autocorrelation function of the squared log returns are plotted in Fig. 4.15. The sample autocorrelation function (ACF) and partial autocorrelation function (PACF) plots show significant autocorrelation in the squared log returns of Bitcoin. The Ljung-Box test formally supports this claim, the null-hypothesis of no-autocorrelation is rejected for the first 40 lags, see Table 4.4, hence, there is volatility clustering in the data.

The significance of the lags in the PACF and ACF plots shows that both the AR and MA part are needed to capture the behaviour of the mean-process of the log returns correctly. The volatility process is simulated using the GARCH model because the ARMA model assumes a constant variance given past information for the log returns, which is clearly not the case for Bitcoin. The GARCH model corrects the ARMA model for the volatility clustering and fat tails in the data, the fat tails are further corrected by using a t-distribution for the residuals. Therefore, in order to model the log returns accurately, we need both the ARMA and GARCH

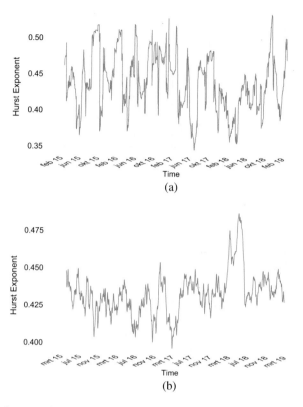

(a)

(b)

Fig. 4.14 The long-run-dependency calculated over a monthly rolling window. (**a**) Returns. (**b**) Volatility

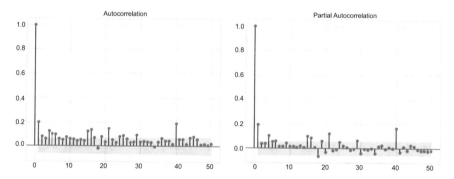

Fig. 4.15 Autocorrelation plots of the squared BTC log returns

Table 4.4 Ljung-Box test for no-autocorrelation on the squared log returns of Bitcoin

Lag	Test statistic	P-value	Lag	Test statistic	P-value
1	45.646	1.417e−11	21	208.671	7.071e−33
2	52.844	3.349e−12	22	211.740	5.646e−33
3	57.372	2.139e−12	23	213.018	1.001e−32
4	74.404	2.663e−15	24	220.160	1.246e−33
5	85.985	4.678e−17	25	229.144	6.771e−35
6	97.105	1.007e−18	26	233.710	2.682e−35
7	101.477	5.342e−19	27	235.012	4.557e−35
8	104.789	4.468e−19	28	236.667	6.526e−35
9	111.441	7.481e−20	29	246.790	2.134e−36
10	115.900	3.428e−20	30	248.146	3.454e−36
11	119.568	2.214e−20	31	249.899	4.633e−36
12	121.922	2.556e−20	32	251.557	6.417e−36
13	125.237	1.850e−20	33	252.606	1.149e−35
14	127.532	2.085e−20	34	252.615	3.205e−35
15	145.599	1.800e−23	35	254.057	4.717e−35
16	167.101	3.186e−27	36	259.481	1.219e−35
17	173.152	6.647e−28	37	261.729	1.249e−35
18	173.444	1.898e−27	38	263.909	1.310e−35
19	180.988	1.977e−28	39	264.241	3.025e−35
20	182.373	3.329e−28	40	308.823	3.136e−43

processes. In other words, the log returns (r_t) follow an ARMA(p_1, q_1)-GARCH(p_2, q_2) process when:

$$r_t = c + \sum_{i=1}^{p_1} \phi_i r_{t-i} + \sum_{i=1}^{q_1} \theta_i \epsilon_{t-i} + \epsilon_t$$

$$\epsilon_t = Z_t \sigma_t$$

$$Z_t \sim t(\nu) \tag{4.3}$$

$$\sigma_t^2 = \omega + \sum_{i=1}^{p_2} \gamma_i \sigma_{t-i}^2 + \sum_{i=1}^{q_2} \psi_i \epsilon_{t-i}^2$$

where ϵ_t denotes the residuals with zero mean of the log returns of Bitcoin at time t, it substitutes the part which cannot be predicted and is generated from the GARCH process with parameters ψ_i and γ_i. Z_t is a noise term of the t-distribution with ν degrees of freedom. The mean-model is predicted by an ARMA-model, where ϕ_i are the autoregressive (AR) coefficients, θ_i are the moving average (MA) coefficients and c represents the mean. All the parameters need to be estimated based on the data of the log returns.

Table 4.5 AIC/BIC values
of the ARMA(p_1, q_1) fit on
the log returns of Bitcoin

(p_1, q_1)	AIC	BIC
(2, 2)	5925.662	5950.450
(3, 3)	5927.039	5961.742
(3, 2)	5927.602	5957.347
(2, 3)	5927.611	5957.356
(4, 2)	5929.202	5963.905

Table 4.6 Model results of an ARMA fit to the log returns of Bitcoin

	Coef.	Std. err.	Z	P-value	95.0% Conf. int.
ϕ_1	−1.694	0.009	−185.838	0.000	[−1.712, −1.676]
ϕ_2	−0.983	0.010	−102.622	0.000	[−1.002, −0.964]
θ_1	1.702	0.011	148.727	0.000	[1.680, 1.725]
θ_2	0.980	0.011	90.315	0.000	[0.959, 1.001]
c	0.0016	0.000	40.906	0.000	[0.002, 0.002]

Coef. denotes the fitted coefficients, Z is the test statistic and the p-value gives an interpretation of the significance of the coefficient and the last column gives a 95% confidence interval of the coefficients

First, we fit an ARMA(p_1, q_1) model, the best combination of (p_1, q_1) is chosen according to two information criteria, namely the Akaike (AIC) and Bayesian (BIC) information criteria. Table 4.5 depicts the different AIC- and BIC-values of several combinations of (p_1, q_1), it is clear that ARMA(2,2) offers simultaneously the lowest AIC- and BIC-value and therefore this model is chosen to model the mean process of the log returns.

Table 4.6 gives the specifications of the fitted ARMA(2,2) model; coef denotes the fitted coefficients, Z is the test statistic and the p-value gives an interpretation of the significance of the coefficient and the last column gives a 95% confidence interval of the coefficients. Note that all p-values for the coefficients are smaller than 0.05, hence the coefficients are deemed significant to predict the mean model. The standard error is 1% for the auto regressive part, this indicates that the Bitcoin log returns are strongly mean-reverting.

The squared residuals of the fitted ARMA(2,2) model still display autocorrelation, see Fig. 4.16. In fact, this is what we expect because the GARCH-model needs to be used on the residuals. On first sight, the mean process is accurately modeled according to the test data as displayed in Fig. 4.17. However, we can only be certain the ARMA-GARCH model is a good fit after the GARCH model is specified.

We will now simultaneously fit the ARMA and GARCH models on the data, as explained in Eq. (4.3). We will scale the returns by 100 before estimating the ARMA-GARCH model, this helps the optimizer to converge, since the scale of the volatility intercept is much closer to the scale of the other parameters in the model. Moreover, Z_t is given a student t-distribution due to the heavy tails of the distribution of Bitcoin log returns. The lowest AIC/BIC values across different combinations of (p_2, q_2) determine the best fitting GARCH model. Table 4.7 depicts

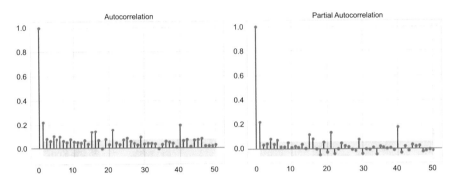

Fig. 4.16 Autocorrelation plot of the model residuals squared

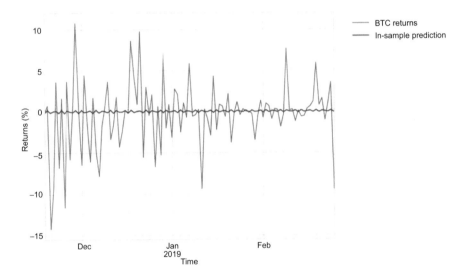

Fig. 4.17 Mean-process of the model

Table 4.7 AIC/BIC values of the best fitting GARCH(p_2, q_2) model on the log returns of Bitcoin

(p_2, q_2)	AIC	BIC
(1, 3)	5365.070	5394.815
(2, 3)	5366.980	5401.682
(3, 3)	5368.980	5408.640
(1, 2)	5369.999	5394.786
(2, 2)	5371.999	5401.744

Table 4.8 ARMA(2,2)-
GARCH(1,3) model results
on the log returns of Bitcoin

	Coef.	Std. err.	T-value	P-value
c	0.217	0.056	3.84794	0.000
ϕ_1	−0.005	0.003	−1.525	0.127
ϕ_2	−0.991	0.003	−329.674	0.000
θ_1	−0.001	0.003	−0.368	0.713
θ_2	0.993	0.000	5118.236	0.000
ω	0.225	0.113	1.989	0.047
γ_1	0.267	0.042	6.309	0.000
ψ_1	0.123	0.108	1.142	0.254
ψ_2	0.193	0.098	1.961	0.049
ψ_3	0.416	0.097	4.301	0.000
ν	3.340	0.259	12.884	0.000

the AIC/BIC values for these different combinations, as a result, $(p_2, q_2) = (1, 3)$
provides the best combined AIC/BIC score.

Table 4.8 provides the model specifications for the ARMA(2,2)-GARCH(1,3)
model fitted to the data. The ARMA-GARCH model in full (with the scaled returns
r_t^* for $t = 1 \ldots n$) is

$$r_t^* = 0.217 - 0.005r_{t-1}^* - 0.991r_{t-2}^* - 0.001\epsilon_{t-1} + 0.993\epsilon_{t-2} + \epsilon_t$$

$$\epsilon_t = Z_t\sigma_t \text{ where } Z_t \sim t(3.340) \tag{4.4}$$

$$\sigma_t^2 = 0.225 + 0.267\sigma_{t-1}^2 + 0.123\epsilon_{t-1}^2 + 0.193\epsilon_{t-2}^2 + 0.416\epsilon_{t-3}^2.$$

Notice that the parameters of the final ARMA(2,2) part are different from the
previously specified ARMA(2,2) model because the final model is constructed by
simultaneously fitting the ARMA and GARCH part. The sum of the α and ψ
coefficients are close to 1, which indicates a high degree of volatility persistence.
The estimated GARCH coefficients are all significant except for ψ_1 at a 5%
significance level. ϕ_1 and θ_1 are both not-significant at a 5% significance level and
both coefficients are also close to zero, this means that they do not differ significantly
from zero. Note that θ_2 is highly significant and large, this confirms the hypothesis
of the autocorrelation structure.

Table 4.9 represents the Ljunx-Box (LB) test for the squared residuals and the
Arch LM (ARCH) test on the standardized residuals, with null-hypothesis':

$$H_0(LB) : \text{ The residuals have no-autocorrelation}$$

$$H_0(ARCH) : \text{ The residuals have no-arch effect.}$$

The p-value for the two tests is higher than 0.05, hence it is not possible to reject
the null-hypothesis in both cases. The chosen model is probably appropriate for
the Bitcoin log returns since there are no arch-effects left and the residuals do not
exhibit volatility clustering.

Table 4.9 The Ljunx-Box (LB) test for the squared residuals and the Arch LM (ARCH) test on the standardized residuals, which respectively test the null-hypothesis of no-autocorrelation and no-arch effects

	Test statistic	P-value
ARCH(5)	0.07081	0.7902
LB	0.03135	0.8595

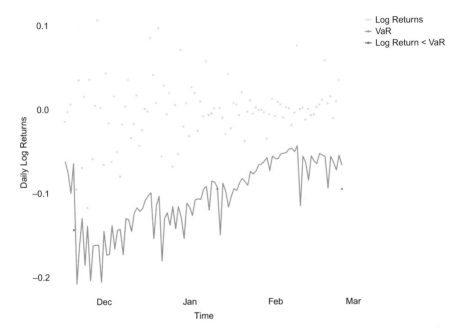

Fig. 4.18 95% Value-at-Risk forecast using a rolling window of 100 days

The ARMA-GARCH model also allows for Value-at-Risk (VaR) prediction, which can be calculated as

$$VaR(\alpha) = \hat{\mu}_{t+1} + \hat{\sigma}_{t+1} t_{\alpha}(\nu),$$

where α denotes the confidence level, $\hat{\mu}_{t+1}$ and $\hat{\sigma}_{t+1}$ need to be estimated based on the ARMA-GARCH model and $t_{\alpha}(\nu)$ is the α-quantile of a t-distribution with ν degrees of freedom. The model is able to forecast the VaR, we use a one step ahead forecast with a rolling window of 100 days. Figure 4.18 provides an overview of the 95% VaR forecast of the test dataset, the actual returns and the breaches of the VaR are shown on the graph. There are only 3 breaches on 100 days. We use the conditional (Christoffersen) and the unconditional (Kupiec) test to check the accuracy of the VaR, with null-hypothesis:

$$H_0 : \text{Correct exceedances of the VaR limit.}$$

Table 4.10 Results of the
conditional and unconditional
test for exceedances on the
log returns of Bitcoin

	Test statistic	P-value
Kupiec	1.616	0.204
Christoffersen	2.841	0.242

Table 4.10 shows the outcome of both the conditional and unconditional test for exceedances, both tests have a *p*-value bigger than the significance level (0.05), therefore, we can say that the VaR limit is accurate.

4.7   Market Efficiency

In this section, we will test the efficiency of the cryptocurrency market. The efficient market hypothesis developed by Malkiel and Fama (1970) and offers three types of market efficiency. The weak form of market efficiency says that past returns cannot be used to predict the future. Due to the erratic behaviour of Bitcoin, it is most likely that if cryptocurrency markets are efficient, they will uphold the weak form.

Urquhart (2016) finds that Bitcoin is not weakly efficient, when the hypothesis is tested on a sample from 2013 until mid 2016. We test for the weak form of market efficiency on the logreturns of Bitcoin from 2015 until September 2019. We will follow the approach of Urquhart to test if the Bitcoin market is weakly efficient. The Runs test (Wald and Wolfowitz 1940) and the Bartels test (Bartels 1982) check if the returns are independent as null hypothesis. The BDS test (Broock et al. 1996) is a popular non-parametric test for serial dependence where the null-hypothesis states that the returns are i.i.d. and the alternate hypothesis tells that the model is misspecified. The AVR test (Choi 1999) determines if the returns are performing a random walk and the variance of price difference of order q is p times the variance of the price difference (p and q are determined based on the data). The Dominguez and Lobato test (DL test) (Dominguez and Lobato 2003) has as null hypothesis that the returns follow a martingale difference process. The *p*-values of each test are depicted in Table 4.11.

From Table 4.11 one can see that the null-hypothesis of independence cannot be rejected by both the Bartels and Runs test on a 5% significance level. The BDS test rejects the fact that returns are i.i.d. on a 5% significance level. The null hypothesis of the DL test cannot be rejected, therefore the returns might follow a martingale difference process, also the AVR test cannot be rejected. The tests thus suggest that the returns of Bitcoin exhibit weak efficiency, which is also a conclusion of Wei

Table 4.11 *P*-values of the test results of the weak form of market efficiency

Runs test	Bartels test	BDS test	AVR test	DL test-C_p	DL test-K_p
0.09	0.68	0.00	0.65	0.22	0.24

(2018). Moreover, according to Wei (2018), liquidity plays a role in the market efficiency and return predictability.

References

Alexander C, Dakos M (2019) A critical investigation of cryptocurrency data and analysis. Technical report

Bacon C (2008) Practical portfolio performance measurement and attribution, vol 546, 2nd edn.. John Wiley & Sons, Hoboken

Bartels R (1982) The rank version of von Neumann's ratio test for randomness. J Am Stat Assoc 77(377):40–46

Broock WA, Scheinkman JA, Dechert WD, LeBaron B (1996) A test for independence based on the correlation dimension. Econ Rev 15(3):197–235

Burniske C, White A (2017) Bitcoin: ringing the bell for a new asset class. Technical report

Choi I (1999) Testing the random walk hypothesis for real exchange rates. J Appl Econ 14(3):293–308

Dominguez MA, Lobato IN (2003) Testing the martingale difference hypothesis. Econ Rev 22(4):351–377

Kenton W (2018) Flash crash. https://www.investopedia.com/terms/f/flash-crash.asp

Malkiel BG, Fama EF (1970) Efficient capital markets: a review of theory and empirical work. J Financ 25(2):383–417

SFox (2019) A Short History of Bitcoin Flash Crashes. https://blog.sfox.com/bitcoin-flash-crash-short-history-cbcb0d6bc6a5

Urquhart A (2016) The inefficiency of Bitcoin. Econ Lett 148:80–82

Wald A, Wolfowitz J (1940) On a test whether two samples are from the same population. Ann Math Stat 11(2):147–162

Wei WC (2018) Liquidity and market efficiency in cryptocurrencies. Econ Lett 168:21–24

Chapter 5
Futures and Options on Cryptocurrencies

Abstract In Chap. 4, we have extensively discussed the characteristics of cryptocurrency prices. However, there is a whole other data-class we have not yet discussed, namely derivative contracts. On the 24th of July 2017, the U.S. Commodity Futures Trading Commission approved to clear and settle derivatives on cryptocurrencies. Only 3 months later, LedgerX started trading the first options on Bitcoin and another 2 months later, two regulated exchanges got the authorization to trade BTC futures, namely the CBOE futures exchange and the Chicago Mercantile exchange. This chapter investigates the available future and option data of cryptocurrencies to get a better understanding of the cryptocurrency derivatives market.

Keywords Futures · Options · Derivatives · Implied volatility · Volatility smiles · Stochastic volatility · VIX-based volatility

5.1 Futures

Futures are contracts to buy or sell an asset at a predefined time against a predefined price. The number of exchanges who provide futures on cryptocurrencies are scarce. Moreover, one of the only regulated (normal) exchanges, namely the Chicago Board Options Exchange (CBOE), stopped offering futures in March 2019 for an undefined period of time (Hochstein 2019). The listed contracts in March 2019 stayed available for trading until they expired in June 2019. The Chicago Mercantile Exchange Group (CME) is another example of a regulated exchange that offers futures on cryptocurrencies. Some cryptocurrency exchanges, like Kraken, Deribit and Okex, also offer futures on their exchange.

Table 5.1 Symbols used in Eqs. (5.1) and (5.2)

WM_i	Weighted median trade price of time interval i
$p_{i,j}$	The price of the jth price/size trade pair of time interval i
$s_{i,j}$	The size of the jth price/size trade pair of time interval i
I_i	The number of transactions during time interval i

5.1.1 Chicago Mercantile Exchange Futures

The Chicago Mercantile Exchange (CME Group 2019a) is one of the few regulated derivative exchanges who also offers futures on Bitcoin. These futures are traded 23 h a day on business days (from 00h00 to 23h00 GMT) and the contract unit is 5 Bitcoins. The contract expires on the last Friday of the fourth month at 16 h GMT. If that day is not a business day in both the United Kingdom and the United States of America, then the contract terminates the proceeding business day.

The CME calculates its future prices based on a Bitcoin/USD reference rate (BRR) (CME Group 2019b). This rate is calculated once a day by recording the trade flow of all the major spot exchanges. The recording takes place every day starting at 16 h during a 1 h window. The transactions in this window are then split up into 5 min time-interval groups, afterwards the weighted is calculated for every group. Finally, the BRR is equal to the average of all the weighted medians of that particular day, as shown in Eq. (5.1) and the calculation of the weighted median is shown in Eq. (5.2), where the meaning of each symbol is represented in Table 5.1.

$$BRR = \frac{\sum_{i=1}^{12} WM_i}{12} \tag{5.1}$$

with

$$WM_i = \begin{cases} p_{i,j}, & \text{if } \sum_{k=1}^{j-1} s_{i,k} < \frac{\sum_{k=1}^{I_i} s_{i,k}}{2} \ \& \ \sum_{k=j+1}^{I_i} s_{i,k} \leq \frac{\sum_{k=1}^{I_i} s_{i,k}}{2} \\ p_{i,1}, & \text{if } s_{i,1} > \frac{\sum_{k=1}^{I_i} s_{i,k}}{2} \\ \frac{p_{i,j}+p_{i,j+1}}{2}, & \text{if } \sum_{k=j+1}^{I_i} s_{i,k} = \frac{\sum_{k=1}^{I_i} s_{i,k}}{2} \end{cases} \tag{5.2}$$

The strategy described in Eqs. (5.1) and (5.2) aims at protecting the reference rate against price anomalies.

5.1.2 Cryptocurrency Exchanges Trading Futures

The futures on Deribit trade 24 h a day, and 7 days a week and are cash settled (BTC) rather than a physical delivery of the cryptocurrency (Deribit 2019). In other words, once the contracts expire, there will only be a transfer of loss/gain between both parties of the future contract. The expiration time is always at 8 h GMT on the last Friday of the expiration month. There exist two quarterly features (expiry last Friday of March, June, September and December). Deribit offers futures on both ETH/USD and BTC/USD. The loss/gain of the contract is calculated based on the time weighted average of the Deribit Ethereum or Bitcoin price index, where the time-interval spans the last half hour before expiration.

Kraken offers futures with an expiration of either monthly, quarterly or perpetual on ETH/USD, XRP/USD, BTC/USD, LTC/USD, BCH/USD and XRP/BTC (Kraken 2019).

A perpetual future has no expiration date, the holder of the contract can decide when the contract expires. The price of a perpetual future lies very close to the spot price of the underlying. In order to keep the price of the future close to the spot price, a funding mechanism is applied. In this mechanism the funding rate is paid between the long position and short position holders, when the funding rate is positive the long position holder pays the short position holder the rate and when it is negative the payments happens in the other direction.

The contract size is always 1 USD for the first five types of contracts and 1 XRP for the last contract of Kraken. Moreover, the first five types are inverse futures, while a future on XRP/BTC is a vanilla future. An inverse future settles futures in the particular cryptocurrency, while regular future markets settle in USD. As a result, Kraken is able to offer traders exposure to "cryptocurrency"/"fiat" pairs without accepting fiat money deposits. Basically, the profit/loss is then calculated as:

$$\text{profit/loss} = \left(\frac{1}{P_{Close}} - \frac{1}{P_{Start}} \right) \cdot \#\text{contracts}.$$

Okex (2019) is another cryptocurrency exchange which offers futures on BTC/USD, ETH/USD, LTC/USD, ETC/USD, XRP/USD, EOS/USD, BCH/USD, BSV/USD and TRX/USD. The futures are either quarterly, bi-weekly, weekly or perpetual expiration. The cryptocurrency exchanges traded its futures 24 h a day and 7 days a week.

5.2   Options on Cryptocurrencies

Options are contracts that give the buyer the right to buy (call option) or sell (put option) an asset at a predetermined price (called the strike price) and time in the future (the maturity time). Note that the buyer has the right to act not the obligation as is the case in futures. We will consider European style options, these options are exercised at maturity. An option can be cash settled or physically settled. In a cash settled option (in cryptocurrency) there is no physical delivery of the underlying, only the possible profit is paid out. A physically settled option, on the other hand, does transfer the underlying to the buyer of the contract for the predefined price.

Deribit (2019), Quedex (2019) and LedgerX (2019) are cryptocurrency exchanges that offer European style options which expire on Fridays at 20h GMT. Deribit offers options on Ethereum and Bitcoin, while LedgerX and Quedex only offer options on Bitcoin. Another difference between these exchanges is that Deribit and Quedex are examples of "cash" settled options (in BTC or ETH depending on the exchange) while LedgerX is an example of a physically settled option exchange. The contract size on all of the exchanges is always one unit of the underlying cryptocurrency.

The most well-know option pricing model is the Black and Scholes model with stock price process $(S_t)_{t \geq 0}$:

$$\frac{dS_t}{S_t} = r dt + \sigma dW_t \tag{5.3}$$

with σ the volatility, $(W_t)_{t \geq 0}$ a standard Brownian motion and r the riskfree rate. We recall the Black and Scholes formula's for European call and put options:

$$C(S_0, K) = S_0 N(d_1) - K \exp(-rT) N(d_2),$$
$$P(S_0, K) = K \exp(-rT) N(-d_2) - S_0 N(-d_1),$$
$$d_1 = \frac{\ln(S_0/K) + (r + \sigma^2/2)T}{\sigma \sqrt{T}}, \tag{5.4}$$
$$d_2 = d_1 - \sigma \sqrt{T}$$

where $C(\cdot, \cdot)$ is the call option price, $P(\cdot, \cdot)$ is the put option price, S_0 is current stock (or other underlying) price, K is the strike price, $N(\cdot)$ denotes the normal distribution and T is the maturity. From this formula, one can see that the normal distribution is used to model stock prices. Section 4.5 shows that the normal distribution is not a very good fit for cryptocurrency price data, the distribution of cryptocurrencies has much fatter tails and is more peaked. However, this model is useful to calculate the implied volatility. To find the implied volatility, we enforce that the market price of the option is equal to the model price of the option using the Black and Scholes option pricing model:

$$P_{Model} = P_{Market}.$$

The parameter σ for which the equation holds is called the implied volatility. Note that if the Black and Scholes model would be the correct model, then the implied volatility would be independent of K and T.

5.3   Volatility Measures of Cryptocurrencies

It is a well-known fact that cryptocurrencies are highly volatile. Options on the exchange rate between Bitcoin and the Dollar will provide further evidence that Bitcoin is a volatile asset.

In this section, we will compute the historical volatility by using a weighting scheme, the implied volatility according to the Black and Scholes option pricing model and a VIX-like volatility index based on Bitcoin option data.

First, we will compute the historical volatility based on daily data using Eq. (5.5) with a window size of 30 days,

$$\sigma_t^2 = \frac{b-1}{b+1}\sigma_{t-1}^2 + \frac{2}{b+1}r_{t-1}^2, \tag{5.5}$$

where b denotes the number of days included in calculating the weights, this means that $b = 30$. The weighting scheme ensures that the more recent volatilities have a higher impact in the calculation.

The second approach calculates the implied volatility based on options prices of Bitcoin. The options are cash settled European style options. The underlying price of the option depends on an index which is calculated based on the prices from six major exchanges, namely Itbit, Kraken, Bitfinex, Gemini, Bitstamp and GDAX.

We will look at the implied volatility smiles for different option surfaces of Bitcoin, as shown in Fig. 5.1. The implied volatilities in this figure are constructed by using out-of-the-money options for each maturity with calculation date the 14th of June 2019. The implied volatility curves in this figure have the shape of a smile and the smile is more or less symmetric around the spot price. These figures also immediately show that the (implied) volatility of Bitcoin is high and not constant over the different strike prices. As a result, this shows that the Black and Scholes model may not be suitable for cryptocurrency data.

Finally, another volatility measure obtained from option prices based on the VIX white-paper by CBOE (2019) is calculated (see also Alexander and Imeraj (2019) for a similar approach and further information). In this approach, we calculate the volatility index as introduced by CBOE on option data of Bitcoin.

Next, we illustrate the above described volatility measures by using historical data. The option data used starts from mid February 2019 to July 2019 and the historical data dates back to 2015. Moreover, we assume an interest rate of 2.5%. The implied volatility shown in Fig. 5.3, is constructed by taking all the at-the-money options with a time until maturity of approximately 3 months into account.

The historical volatility is on average around 68%, as shown in Fig. 5.2, moreover, the historical volatility seems to be mean-reverting. The historical (yearly) volatility even shows peaks of over 150%, this means that the standard deviation of a yearly return of Bitcoin expressed in Dollars has a standard deviation of 150%. One can also see from this figure that the volatility of the volatility is very high. For

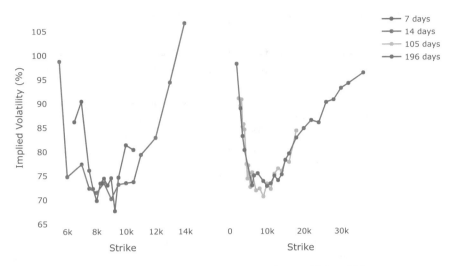

Fig. 5.1 Implied volatility smiles for out-of-the-money options on 14 June 2019

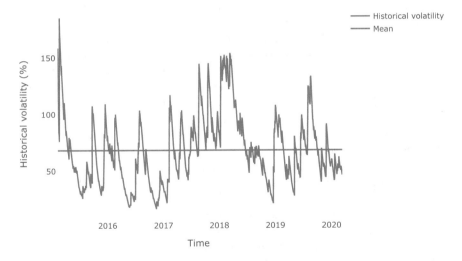

Fig. 5.2 Historical volatility of Bitcoin returns

example in February 2015 the volatility was 187% and it dropped to 23.91% only 4 months later.

Figure 5.3 compares the different volatility measures during the period from February 2019 to July 2019. The volatility has a slightly downward trend until the end of March 2019 according to the different volatility measures. Historical volatility is backward-looking, hence the big jumps on certain days may be corrections of the historical volatility when new data is inserted. The volatility

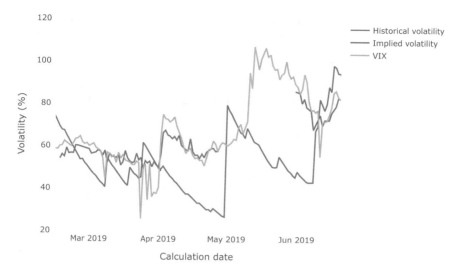

Fig. 5.3   Volatility of Bitcoin according to the different volatility measures

fluctuates around 70% according to all the measures, the implied volatility is rather
steady over time while the VIX-based volatility can be more erratic.

5.4   Stochastic Volatility

This section focuses on the Heston pricing model (Heston 1993) (see also Madan
et al. (2019)). The Heston model defines volatility as a stochastic mean-reverting
process. Recall indeed that in the previous section most of the volatility estimates
showed a lot of fluctuation.

First, we will introduce the Heston model, this model allows us to analyse the
high fluctuations in the volatility through the vol-of-vol parameter (θ). The Heston
model is a variation on the Black and Scholes model where the volatility follows a
stochastic process. The risk neutral price process $(S_t)_{t \geq 0}$ is defined as:

$$\frac{dS_t}{S_t} = r dt + \sqrt{v_t} dW_t$$

where

$$dv_t = \kappa(\eta - v_t)dt + \theta \sqrt{v_t} d\tilde{W}_t$$

with $\kappa > 0$ is the rate of mean reversion, $\eta > 0$ is the long run variance and $v_0 > 0$ is the initial variance. $(W_t)_{t \geq 0}$ and $(\bar{W}_t)_{t \geq 0}$ are two standard Brownian motions with

$$Cov(\mathrm{d}W_t, \mathrm{d}\bar{W}_t) = \rho \mathrm{d}t.$$

where $-1 \geq \rho \geq 1$ is the correlation between the stock and the volatility. In this case, the characteristic function of $\log(S_T)$ is given by

$$\phi_T(u) = \exp(iu \log(S_0))$$
$$\cdot \exp(rTui + \eta\kappa\theta^{-2}((\kappa - \rho\theta ui - d)T - 2\log((1 - g\exp(-dT))(1 - g)^{-1})))$$
$$\cdot \exp(v_0\theta^{-2}(\kappa - \rho\theta iu - d)(1 - \exp(-dT))(1 - g\exp(-dT))^{-1})$$

(5.6)

with

$$d = ((\rho\theta ui - \kappa)^2 - \theta^2(-iu - u^2))^{1/2}$$
$$g = (\kappa - \rho\theta ui - d)(\kappa - \rho\theta ui + d)^{-1}.$$

Vanilla options prices under the Heston model are calculated using the Fast Fourier Transform methodology (Carr and Madan 1999). In this case, the price of a call option with strike K and maturity T is given by:

$$C(K, T) = \frac{\exp(-\alpha \log(K))}{\pi} \int_0^\infty \exp(-iv \log(K))\rho(v)\mathrm{d}v$$

where

$$\rho(v) = \frac{\phi_T(v - (\alpha + 1)i)}{\alpha^2 + \alpha - v^2 + i(2\alpha + 1)v},$$
$$\alpha = 1.5$$

and the characteristic function ϕ_T of the log price process at time T.

The model parameters are determined by minimizing the root mean squared error (RMSE) of the model price with respect to the market price of the option (see Eq. (5.7)).

$$RMSE = \sqrt{\frac{1}{n}\sum_{i=1}^n (P_{Market,i} - P_{Model,i})^2}$$

(5.7)

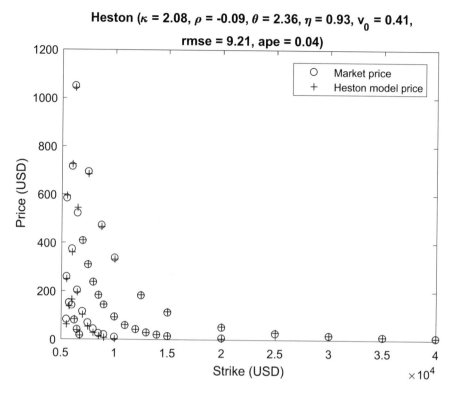

Fig. 5.4 Calibration of the Heston model on 29/06/2018

In order to determine if the chosen model is a good fit, we will apply another measure, namely the average pricing error (APE) (see Eq. (5.8)).

$$APE = \frac{\sum_{i=1}^{n} |P_{Market,i} - P_{Model,i}|}{\sum_{i=1}^{n} P_{Market,i}} \tag{5.8}$$

The data consists of European-style vanilla options, these options are cash settled in BTC and have USD as a numeraire. The underlying BTC price index is based on price data from six different exchanges, namely Bitfinex, Kraken, Bitstamp, GDAX, Gemini and Itbit. Moreover, the data is extracted over a period starting from Friday 29 June 2018 until Friday 31 August 2018. During the calibration, only consider out-of-the-money options where both the bid and the ask price are available because out-of-the-money options only have time value and this makes the calibration more robust.

Figure 5.4 depicts the market (o-signs) and model prices (+ signs) of the options, where the model prices are based on the Heston model with optimal parameters. From this figure, one can see that the model provides a reasonable good fit, moreover, the RMSE and APE are respectively only 9.21 and 0.04. The Heston

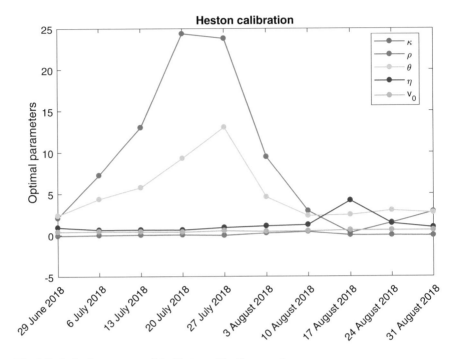

Fig. 5.5 Optimal parameters of the Heston calibration over time

model estimates the initial volatility as 41% and the long run variance as 93%, which immediately supports the claims in the previous section, namely that Bitcoin is highly volatile. Furthermore, the vol-of-vol parameter is 236%, which shows that the volatility of Bitcoin is also high. Hence, not only the volatility is high but also the volatility of the volatility. Figure 5.5 shows how the optimal parameters change per option surface. The vol-of-vol parameter itself also changes over time, reaching a maximum of nearly 1300% at the end of July 2018.

References

Alexander C, Imeraj A (2019) The crypto investor fear gauge and the Bitcoin variance risk premium. Technical report

Carr P, Madan D (1999) Option valuation using the fast Fourier transform. J Comput Financ 2:61–73

CBOE (2019) White paper: CBOE volatility index. Technical report

CME Group (2019) Bitcoin futures. https://www.cmegroup.com/trading/bitcoin-futures.html

CME Group (2019) CME CF cryptocurrency reference rates: Methodology guide. Technical report

Deribit (2019). Deribit. https://www.deribit.com/

Heston SL (1993) A closed-form solution for options with stochastic volatility with applications to bond and currency options. Rev Financ Stud 6(2):327–343

Hochstein M (2019) CBOE exchange puts brakes on Bitcoin futures listing

Kraken (2019) Bitcoin futures trading, cryptocurrency futures market. https://www.kraken.com/en-us/features/futures

Ledger Holdings Inc. (2019) Ledgerx. https://www.ledgerx.com/

Madan DB, Reyners S, Schoutens W (2019) Advanced model calibration on Bitcoin options. Dig Finan 1:117–137

Okex.com (2019) Futures. https://www.okex.com/future/trade

Quedex Ltd. (2019) Options specifications quedex. https://quedex.net/doc/options

Chapter 6
Portfolio Management

Abstract Cryptocurrencies are often seen as an excellent diversifier for a portfolio of securities. This chapter compares several simulated portfolio consistent of a gold index, mid and large cap ETF, bonds and technological asset with a portfolio of the same assets and Bitcoin. The weights of the assets are determined according to different portfolio mechanisms, namely the Markowitz mean variance portfolio, safety first portfolio based on expected shortfall, an equivalent risk contribution portfolio and an equal weighted portfolio. The performance of the different portfolios are compared by using several performance measures.

Keywords Markowitz portfolio · Equivalent risk contribution portfolio · Safety first portfolio · Rebalance · Maximal drawdown · Volatility · Sharpe ratio · Recovery rate · Turnover rate · Cumulative return

This chapter is dedicated to compare the portfolios investing in a series of traditional assets with and without Bitcoin according to different portfolio construction mechanisms. The analysis includes Markowitz' mean-variance portfolios (Markowitz 1952), a safety first portfolio and an equivalent risk contribution portfolio. Hu et al. (2019) perform an analysis on a larger range of cryptocurrencies, they include Bitcoin, Ethereum, Ripple, Litecoin, Bitcoin Cash and Binance coin in their portfolio. In contrast to our research, they compare a portfolio solely consisting of the aforementioned cryptocurrencies to a benchmark ETF, namely S&P 500. They conclude that a cryptocurrency portfolio outperforms the S&P 500 on cumulative return and depending on the construction mechanism also on the Sharpe ratio. However, these portfolios are more volatile and have a higher maximal drawdown.

Markowitz (1952) uses the mean return and the standard deviation of a portfolio to select the optimal combination of securities. The mean return is estimated as

$$\mu(\boldsymbol{w}) = \sum_{i=1}^{N} w_i \cdot \mu_i.$$

μ_i denotes the historical mean return of security i in the portfolio. The minimization problem of a Markowitz mean-variance portfolio with corresponding constraints is:

$$\min_{\boldsymbol{w}} \sqrt{\boldsymbol{w}\Sigma\boldsymbol{w}} - \eta\mu(\boldsymbol{w})$$

$$\text{s.t.} \sum_{i=1}^{N} w_i = 1, \, w_i \geq 0 \text{ for } i = 1, \ldots, N \ \& \ \mu(\boldsymbol{w}) \geq \eta, \tag{6.1}$$

where η is a number between 0 and 1 and it denotes the risk tolerance. Σ is the covariance matrix of the returns. The minimal variance portfolio chooses $\eta = 0$. In this book, the mean-variance portfolio sets η equal to 0.5.

The safety first portfolio is constructed by minimizing the expected shortfall with respect to the budget constraint ($\sum_{i=1}^{N} w_i = 1$) and the no-short-selling constraint ($w_i \geq 0$ for $i = 1, \ldots, N$). Expected shortfall can be calculated using:

$$ES_\alpha(L) = \frac{1}{1-\alpha} \cdot [E(L \cdot \mathbb{1}(L \geq Q_\alpha)) + Q_\alpha \cdot (1 - \alpha - P(L \geq Q_\alpha))], \tag{6.2}$$

where $\alpha \in (0, 1)$ is the confidence level, $L \in \mathcal{L}^1(\Omega, F, P)$ and Q_α is the α-quantile corresponding to the distribution function of the returns. The term $Q_\alpha \cdot (1 - \alpha - P(L \geq Q_\alpha))$ can be interpreted as a correction term when $P(L > Q_\alpha) > 1 - \alpha$. If $P(L > Q_\alpha) = 1 - \alpha$, as is the case for continuous distribution, then the last term of Eq. (6.2) drops and the equation of expected shortfall reduces to

$$ES_\alpha(L) = E[L|L \geq Q_\alpha] = E[L|L \geq VaR_\alpha]. \tag{6.3}$$

The equivalent risk contribution portfolio solves the following minimization problem:

$$\min_{\boldsymbol{w}} \sum_{i=1}^{N} \sum_{j=1}^{N} (w_i \cdot (\Sigma\boldsymbol{w})_i - w_j \cdot (\Sigma\boldsymbol{w})_j)^2$$

$$\sum_{i=1}^{N} w_i = 1 \ \& \ w_i \geq 0 \text{ for } i = 1, \ldots, N. \tag{6.4}$$

We will analyse the performance of the portfolios according to different performance measures:

1. Cumulative return of the portfolio
2. Volatility of the portfolio

3. Sharpe ratio (Sharpe 1994): a measure for the risk adjusted (calculated in volatility) return.

$$\text{Sharpe}(T) = \frac{\bar{R}(T) - R_f}{\sigma_p} \tag{6.5}$$

where $\bar{R}(T)$ is the mean return of the portfolio, R_f is the risk-free rate and σ_p is the standard deviation of the portfolio on time-frame $[0, T]$.

4. Maximal drawdown (Chekhlov et al. 2011): the maximal observed loss measured from peak to bottom in a specified period $[0, T]$.

$$\sup_{t \in [0,T]} [\sup_{s \in [0,t]} (S_s - S_t)] \tag{6.6}$$

where S_k is the stockprice at time k.

5. Recovery rate: the number of days it takes for the portfolio to recover from a drop.

6. Turnover rate (Demiguel et al. 2009): how much the weights of the assets change each time we rebalance.

$$\text{Turnover} = \frac{1}{\text{trading days}} \sum_{t=1}^{T} \sum_{j=1}^{N} |w_{j,t} - w_{j,t-1}| \tag{6.7}$$

6.1 Portfolio Simulation

This section will compare portfolios investing in a series of traditional assets with and without Bitcoin. The data for the portfolio ranges from 01/01/2015 until 01/07/2019 and includes 6 different assets namely Bitcoin (BTC), Gold index (XAU idnex), Vanguard mid and large cap ETF fund (VO resp. VV), iShares USD Corp Bond UCITS ETF (LQD) and iShares Expanded Tech-Software Sector ETF (IGV). Table 6.1 provides an overview of the main statistics of the log returns of these assets. Bitcoin has the highest volatility and maximal value and it has the lowest minimal value. All the assets, except for Bitcoin, have an almost zero log returns as median. On average each of the assets has positive returns.

The portfolio will be set up on a random start date chosen in the selected time period and will exist for 250 trading days (1 year). Each month the weights in the asset will be rebalanced, assuming that rebalancing can be done without any cost.

Figure 6.1 shows the simulated value of portfolio on 100 different random starting dates, according to the different portfolio measures. The green lines represents portfolios which end up with a value larger than the invested amount and the red portfolios have an end value smaller than the initial investment of 1000 USD. From this figure, it is clear that portfolios that include Bitcoin end up in a

Table 6.1 Description of the log returns of the assets in the portfolio

	IGV	LQD	VO	VV	Xau index	BTC
Mean	0.000509	0.000012	0.000176	0.000210	0.000044	0.002043
Std.	0.010293	0.002572	0.007428	0.007146	0.019617	0.038959
Min	−0.067626	−0.014803	−0.042861	−0.041908	−0.108440	−0.237570
25%	−0.001476	−0.000703	−0.001702	−0.001117	−0.006105	−0.010648
50%	0.000000	0.000000	0.000000	0.000000	0.000000	0.002214
75%	0.004568	0.001177	0.002797	0.002059	0.005936	0.017399
Max	0.063108	0.012228	0.045438	0.048246	0.099191	0.225119

larger range of final values for all portfolio mechanisms. For the mean variance and safety first portfolio, there are more portfolios who end up in green with BTC. The opposite is true for the minimal variance portfolio.

Figure 6.2 shows the average performance over the different simulated portfolios, where the portfolios in Fig. 6.2a include Bitcoin and the portfolios in Fig. 6.2b do not. From both figures, it is immediately clear the mean-variance portfolio poses the biggest risk in terms of drawdown and volatility, however, on average it also gives the highest cumulative return. The Sharpe ratio for the non-Bitcoin portfolio is the highest.

The performance of the minimal variance and safety first portfolios resemble each other in maximal drawdown, volatility and cumulative return. The difference between both portfolios can be seen in Sharpe ratio, recovery time and turnover rate. The safety first portfolio needs a longer time to recover from a drop. The Sharpe ratio of the safety first portfolio is higher when the portfolio also consist of BTC. The turnover rate of the safety first portfolio is also higher. Note that the two portfolios have on average, over all the performance measures, roughly the same values if the portfolio does not include Bitcoin. The differences between these portfolios are more pronounced when Bitcoin is included. Both of these portfolios can be deemed as the least risky in terms of volatility and drawdown. However, it takes these portfolios often the longest to recover from a drop.

The equivalent risk contribution portfolio has a rather low maximal drawdown and volatility, while it offers a high Sharpe ratio and a low recovery time for both portfolio mixes. The equally weighted portfolio offers a high cumulative return and Sharpe ratio. However, the volatility and drawdown are higher than the equivalent risk contribution, minimal variance and safety first portfolios. This portfolio type has the advantage that the weights in the assets never change and therefore there is no rebalancing cost.

If we compare the portfolio measures with Bitcoin in Fig. 6.2a to the those without Bitcoin. Then, the maximal drawdown and volatility are roughly unchanged for the minimal variance, safety first, equivalent risk contribution (ERC) and equal weighted portfolios. Contrary to the cumulative return, which has doubled for the minimal variance and safety first portfolio. Moreover, it has six folded for the equivalent risk contribution and equal weighted portfolio. The recovery time is less

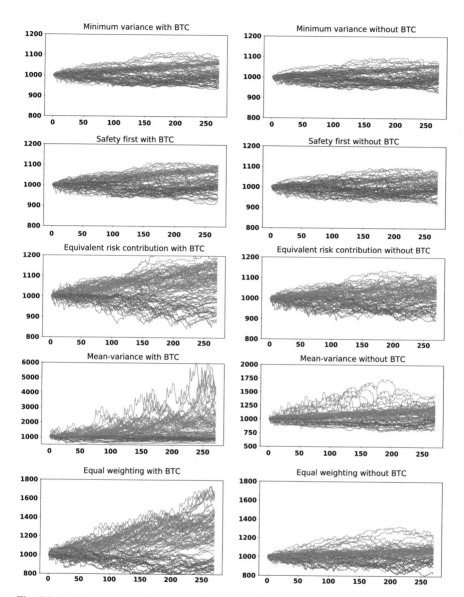

Fig. 6.1 Portfolio value simulation according to the different portfolios construction mechanisms with an initial investment of 1000 USD. Red lines indicate that the portfolio ends up below 1000 USD after 1 year. The y-axis shows the value of the portfolio and the x-axis are the days

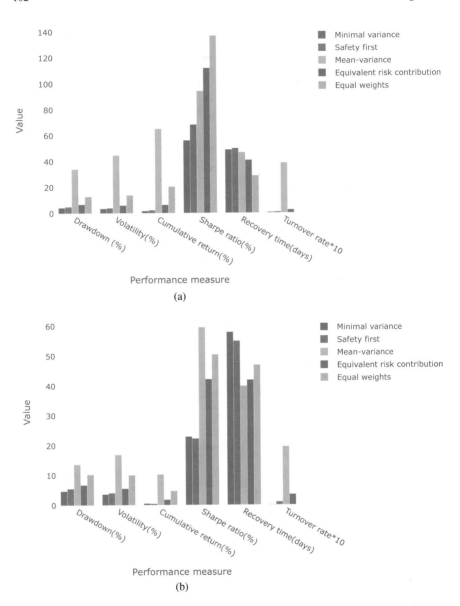

Fig. 6.2 Average performance measures of the portfolio with investment of 1000 USD. The first 4 performance measures are expressed in percentage and the last performance measure in days till recovery. (**a**) Portfolio with BTC. (**b**) Portfolio without BTC

for the portfolios with Bitcoin, for the ERC and equal weighted portfolio it has even halved. The Sharpe ratio has approximately tripled in all these portfolios and the turnover rate remained constant.

The mean-variance portfolio is different from the other portfolio mechanisms, here the maximal drawdown, volatility and turnover rate all doubled. The cumulative return has six folded, the Sharpe ratio is 1.5 times larger and the turnover rate has doubled. This portfolio offers a better return at the cost of maximal drawdown, volatility and turnover rate. When one constructs a portfolio according to this measure, the return can be very profitable, however, it comes at the cost of a highly volatile portfolio.

A final note of caution, Bitcoin has seen a rally in value over the past years. Therefore, it is not surprising that including Bitcoin can alter the return of the portfolio for the better.

References

Chekhlov A, Uryasev S, Zabarankin M (2011) Drawdown measure in portfolio optimization. Int J Theor Appl Financ 8(1):13–58

Demiguel V, Garlappi L, Nogales F, Uppal R (2009) A generalized approach to portfolio optimization: improving performance by constraining portfolio norms. Manag Sci 55:798–812

Hu Y, Rache S, Fabozzi F (2019) Modelling crypto asset price dynamics, optimal crypto portfolio, and crypto option valuation. arXiv.org

Markowitz H (1952) Portfolio selection. J Financ 7(1):77–91

Sharpe WF (1994) The sharpe ratio. J Portf Manag 21:49–58

Chapter 7
Further Related Work

Abstract Cryptocurrencies are highly volatile assets, part of their volatility can be explained by news articles and more particularly the tone or sentiment of this articles. Moreover, more trading in cryptocurrencies also influences the amount of news. Furthermore, cryptocurrencies have a lot of periods where the price significantly deviates from its fundamental value, also known as bubbles. In this chapter, the reader can find an overview of news impact on cryptocurrencies and a thorough overview of bubble behavior.

Keywords Google trends data · New impact · Bubble analysis · Spillover effect · Investor attention

In this section we will discuss other relevant contributions to risk management in cryptocurrencies. We will start by discussing the impact of investor attention on Bitcoin (and other cryptocurrencies) price dynamics. Afterwards, we will elaborate on price bubbles in cryptocurrencies.

7.1 News Impact on Cryptocurrencies

Crytocurrencies have received a great deal of media and investor attention due to their peer-to-peer payment system, no government interference, low transaction costs and high volatile nature, which allowed investors to make high profits on the short term. In the literature there have been numerous studies which show that Bitcoin prices, interest and trading volume are influenced by the amount of "news" on Bitcoin (Cretarola et al. 2020; Shen et al. 2018; Kristoufek 2013; Urquhart 2018; Yelowitz and Wilson 2015). The term "news" does not only cover news articles but also tweets, searches on the internet, ...

Yelowitz and Wilson (2015) have conducted a comparison between Google trends data and the interest in Bitcoin with respect to the different types of users. It turns out that computer programming enthusiasts' search items and illegal activity

© The Author(s), under exclusive licence to Springer Nature Switzerland AG 2020
E. Van der Auwera et al., *Financial Risk Management for Cryptocurrencies*,
SpringerBriefs in Finance, https://doi.org/10.1007/978-3-030-51093-0_7

search terms are positively correlated with Bitcoin interest. A recent paper by Urquhart (2018) also uses Google trends data and claims that previous days realized volatility and volume of Bitcoin are partly influenced by the attention of investors in Bitcoin.

Shen et al. (2018) analyse the relationship between the volume of Tweets mentioning Bitcoin and its returns, trading volume and volatility. They argue that tweets are a better proxy for interest in Bitcoin since these are used by more informed investors. The results indicate that the number of tweets influences the realized volatility and trading volume in Bitcoin. Kristoufek (2013), on the contrary, suggests a relationship between the prices of Bitcoin and the amount of searches in Google and Wikipedia on Bitcoin. Moreover, he claims that the effect on the price of Bitcoin is different depending on being above or below its trend value.

There are also several authors who use models that incorporate investors attention for modelling the prices of cryptocurrencies or options on cryptocurrencies. Cretarola et al. (2020) use the Google searches for "Bitcoin" as a proxy for investor attention. In fact, they have developed a stochastic model in continuous time for modelling Bitcoin option prices which incorporates investor attention. Philippas et al. (2019) use a process diffusion model to examine if Bitcoin prices have jump behaviour, where the jumps stem from Google trends data or Twitter. Their results suggest that Bitcoin prices are only partially driven by media attention. However, the influence is greater in periods of uncertainty.

7.2 Bubble Analysis

A bubble is defined in the literature as a period where the market value of an asset persistently deviates from its fundamental value which cannot be explained by the underlying economic variables (Diba and Grossman 1988). Bitcoin has often risen very sharply, only to fall in value short after. For example in 2013 Bitcoin climbed as high as $1132.26, only to drop to 60% of its value only several months later. There have been other recordings of sharp rises and falls in the history of Bitcoin, like the rise and fall in 2017–2018.

Many authors have investigated the bubble behaviour of Bitcoin, in order to detect reasons of the sharp movements in Bitcoin. Blau (2017) investigated the bubble of 2013 to see if speculation drove this bubble. He concludes that during this period speculative trading was not particularly high. Moreover, there was no multi- or uni-variate relationship between the volatility of Bitcoin and speculative trading.

Cheah and Fry (2015) take the bubble analysis even further and contradict the hypothesis of Blau. In 2015, they claim that Bitcoin is prone to speculative bubbles and the bubble component in Bitcoin is substantial. Moreover, they also determine the fundamental value of Bitcoin. Fry and Cheah (2016) uncover the existence of a spillover effect from Ripple to Bitcoin that exacerbates the price falls in Bitcoin.

They also find evidence that certain events have an impact on the market, like the technical glitch in Bitcoin software in March 2013.

The above mentioned literature investigates bubbles at certain points in time. More recently, Cretarola and Figà-Talamanca 2019 use strict local martingale theory of financial bubbles to define a bubble detection criterion. Moreover, they take use Google searches as a market attention factor, since many papers have found a positive relationship between Bitcoin price evolution and market attention (see previous section). Further, Cretarola and Figà-Talamanca 2020 extend the model of Cretarola and Figà-Talamanca 2019 to a continuous time model which accounts for regime changes, bubbles and market attention in Bitcoin and Ethereum. They achieve this by introducing a regime-switching correlation parameter.

References

Blau BM (2017) Price dynamics and speculative trading in bitcoin. Res Int Bus Financ 41:493–499
Cheah E-T, Fry J (2015) Speculative bubbles in bitcoin markets? An empirical investigation into the fundamental value of bitcoin. Econ Lett 130(C):32–36
Cretarola A, Figà-Talamanca G (2019) Detecting bubbles in Bitcoin price dynamics via market exuberance. Ann Oper Res 1–21
Cretarola A, Figà-Talamanca G (2020) Bubble regime identification in an attention-based model for bitcoin and ethereum price dynamics. Econ Lett 191:108831
Cretarola A, Figà-Talamanca G, Patacca M (2020) Market attention and Bitcoin price modeling: Theory, estimation and option pricing. Decisions Econ Finan 43(1):187–228
Diba BT, Grossman HI (1988) Explosive rational bubbles in stock prices? Am Econ Rev 87:520–530
Fry J, Cheah E-T (2016) Negative bubbles and shocks in cryptocurrency markets. Int Rev Financ Anal 47:343–352
Kristoufek L (2013) Bitcoin meets google trends and wikipedia: quantifying the relationship between phenomena of the internet era. Sci Rep 3(3415):12
Philippas D, Rjiba H, Guesmi K, Goutte S (2019) Media attention and bitcoin prices. Financ Res Lett 30:37–43
Shen D, Urquhart A, Wang P (2018) Does twitter predict bitcoin? Econ Lett 174:118–122
Urquhart A (2018) What causes the attention of bitcoin? Econ Lett 166:40–44
Yelowitz A, Wilson M (2015) Characteristics of bitcoin users: an analysis of google search data. Appl Econ Lett 22(13):1030–1036

Part III
Summary and Conclusion

Chapter 8
Conclusion

Abstract Cryptocurrencies and blockchain hit the world like a bomb. Bitcoin is the first successful attempt to create a decentralized, permissionless, pseudonymous cash system. However, no attempt is perfect from the first try and every new invention brings a specific set of new risks with it. The risks can be due to a number of factors, like the underlying mechanism, the typical structure of cryptocurrencies, the lack of regulation and protection or the liquidity. This chapter summarizes all the main risks in cryptocurrencies.

Keywords Cryptocurrencies · Risk management · Quantitative risks · Qualitative risks

Bitcoin is the first cryptocurrency and since its creation there have been many other cryptocurrencies. Some of these cryptocurrencies where a scam from the beginning, other cryptocoins did not become successful enough to continue and some others were able to secure their place in the market, like Monero, Ethereum, EOS, …. New cryptocurrencies either try to tackle a new section of the market, like Ripple and Stellar who are present in the remittance transfer market or they try to get a part of Bitcoin's market as a replacement of fiat currency. In the literature, there is a debate what the definition of cryptocurrencies is, is it a currency, commodity or a security. Each definition has its own set of supporters and until now there is no clear winner. In general, the cryptocurrency market has grown at an exponential rate with new coins issued every day. Nevertheless, Bitcoin remains to have the largest market share. Not only the market itself but also the number of ICOs per year have grown exponentially. Unfortunately, 80% of all ICOs turned out to be scams.

One of the things that prevents the wide-spread adoption is the fact that cryptocurrencies have so many (new) risks. Regularly, a hack on an exchange/wallet is reported. There are also new types of attacks, like the 51% attack, where the blockchain itself is hacked or a selfish miner attack. These hacks can be small, however, they can also lead to enormous losses which can have a negative impact on the whole market and the credibility of cryptocurrencies. The legislative framework is still very unclear, most regulators adopt a 'wait and see' approach and as a

E. Van der Auwera et al., *Financial Risk Management for Cryptocurrencies*,
SpringerBriefs in Finance, https://doi.org/10.1007/978-3-030-51093-0_8

result there is no global framework in place. However, cryptocurrency exchanges do need to adhere "know-your-customer" and anti-money-laundering practices in most countries. Money-laundering is also not uncommon in cryptocurrencies. A lot of money-launderers use cryptocurrencies because they believe cryptocurrencies are anonymous, however, they are only pseudonymous.Therefore, one has come up with new techniques to hide identities such as mixing or using a more identity hiding coin like Monero.

The risks in cryptocurrencies are not only design/operational related, some risks have a quantitative nature. For example, the log returns of cryptocurrencies exhibit fat tails, high peaks and positive skewness. Therefore, we found evidence that the best fitting distribution is the student t distribution. Moreover, the log returns seem to be mean-reverting. It is hard to divide cryptocurrencies in a known asset class since they are almost independent to any of the other asset classes. The rolling correlation between cryptocurrencies and other asset classes remains below 20% in absolute value and as a result cryptocurrencies can be seen as a differentiated risk reducer. Furthermore, different cryptocurrencies are highly correlated to each other and this correlation has become higher over the last few years. In the beginning cryptocurrencies with similar characteristics such as Bitcoin and Litecoin, were highly correlated. Because cryptocurrencies are hardly related to any of the asset classes, we have constructed portfolios with Bitcoin and some major ETFs. It turns out that Bitcoin can augment the return of the portfolio while maintaining the same level of volatility and maximal drawdown in comparison with the same portfolio without Bitcoin according to minimal variance, minimal expected shortfall, equivalent risk contribution and equal weighted portfolio construction mechanisms. A portfolio which includes Bitcoin has on average a lower recovery time and a higher Sharpe ratio.

Probably one of the most well-known risks is the highly volatile nature of cryptocurrencies. In this book there are numerous arguments to support the high volatility claim by using both option and price data of cryptocurrencies. The historical volatility of cryptocurrencies, for example, is on average much higher than "regular" securities with an average of nearly 70%. Moreover, even the volatility of the volatility of cryptocurrencies is high and the price data also exhibits volatility clustering.

The last chapter, provides the reader with some additional research. Cryptocurrency prices are highly influenced by media attention, which can be measured by Google searches, tweets, ... Some authors use media attention to model the prices of cryptocurrencies and find very promising results. Due to the highly volatile nature of Bitcoin, it is no surprise that cryptocurrencies exhibit "price bubbles".

Index

Printed in the United States
By Bookmasters